Lecture Notes in Artificial Intelligence 8557

Subseries of Lecture Notes in Computer Science

LNAI Series Editors

Randy Goebel
University of Alberta, Edmonton, Canada
Yuzuru Tanaka
Hokkaido University, Sapporo, Japan
Wolfgang Wahlster
DFKI and Saarland University, Saarbrücken, Germany

LNAI Founding Series Editor

Joerg Siekmann
DFKI and Saarland University, Saarbrücken, Germany

T0183269

Petra Perner (Ed.)

Advances in Data Mining

Applications and Theoretical Aspects

14th Industrial Conference, ICDM 2014
St. Petersburg, Russia, July 16-20, 2014
Proceedings

 Springer

Volume Editor

Petra Perner
Institute of Computer Vision and Applied Computer Sciences, IBaI
Kohlenstrasse 2
04107 Leipzig, Germany
E-mail: pperner@ibai-institut.de

ISSN 0302-9743 e-ISSN 1611-3349
ISBN 978-3-319-08975-1 e-ISBN 978-3-319-08976-8
DOI 10.1007/978-3-319-08976-8
Springer Cham Heidelberg New York Dordrecht London

Library of Congress Control Number: 2014942574

LNCS Sublibrary: SL 7 – Artificial Intelligence

Typesetting: Camera-ready by author, data conversion by Scientific Publishing Services, Chennai, India

Printed on acid-free paper

Springer is part of Springer Science+Business Media (www.springer.com)

Preface

The 14th event of the Industrial Conference on Data Mining ICDM was held in St. Petersburg (www.data-mining-forum.de) running under the umbrella of the World Congress "The Frontiers in Intelligent Data and Signal Analysis, DSA 2014."

After the peer-review process, we accepted 25 high-quality papers for oral presentation, of which 16 are included in this proceeding book. The topics range from theoretical aspects of data mining to applications of data mining, such as in multimedia data, in marketing, in medicine and agriculture, and in process control, industry and society. Extended versions of selected papers will appear in the*International Journal Transactions on Machine Learning and Data Mining* (www.ibai-publishing.org/journal/mldm).

In all, 10 papers were selected for poster presentations and 6 for industry paper presentations that are published in the *ICDM Poster and Industry Proceeding* by *ibai-publishing* (www.ibai-publishing.org).

In conjunction with ICDM, 4 workshops were run focusing on special hot application-oriented topics in data mining: the Workshop on Case-Based Reasoning (CBR-MD), Data Mining in Marketing (DMM), and I-Business to Manufacturing and Lifesciences (B2ML). All workshop papers are published in the *workshop proceedings* by ibai-publishing house (www.ibai-publishing.org).

A tutorial on Data Mining, a tutorial on Case-Based Reasoning, a tutorial on Intelligent Image Interpretation and Computer Vision in Medicine, Biotechnology, Chemistry & Food Industry, and a tutorial on Standardization in Immunofluorescence were held before the conference.

We were pleased to give out the best paper award for ICDM for the seventh time this year. There are 4 announcements mentioned at www.data-mining-forum.de. The final decision was made by the Best Paper Award Committee based on the presentation by the authors and the discussion with the auditorium. The ceremony took place at the end of the conference. This prize is sponsored by ibai solutions (www.ibai-solutions.de), one of the leading companies in data mining for marketing, Web mining and e-commerce.

The conference was rounded up by an outlook session on new challenging topics in data mining before the Best Paper Award Ceremony.

We would like to thank all reviewers for their highly professional work and their effort in reviewing the papers.

We also thank the members of the Institute of Applied Computer Sciences, Leipzig, Germany (www.ibai-institut.de), who handled the conference as secretariat. We appreciate the help and understanding of the editorial staff at Springer Verlag, and in particular Alfred Hofmann, who supported the publication of these proceedings in the LNAI series.

Last, but not least, we wish to thank all the speakers and participants who contributed to the success of the conference. We hope to see you in 2015 in Hamburg at the next World Congress "The Frontiers in Intelligent Data and Signal Analysis, DSA 2015" (www.worldcongressdsa.com) that combines under its roof the following three events: International Conferences Machine Learning and Data Mining MLDM, the Industrial Conference on Data Mining ICDM, and the International Conference on Mass Data Analysis of Signals and Images in Medicine, Biotechnology, Chemistry and Food Industry MDA.

July 2014   Petra Perner

Industrial Conference on Data Mining, ICDM 2014

Chair

Petra Perner   IBaI Leipzig, Germany

Committee

Ajith Abraham	Machine Intelligence Research Labs, USA
Andrea Ahlemeyer-Stubbe	ENBIS, The Netherlands
Brigitte Bartsch-Spörl	BSR Consulting GmbH, Germany
Orlando Belo	University of Minho, Portugal
Marc Boullé	France Télécom, France
Shirley Coleman	University of Newcastle, UK
Juan M. Corchado	Universidad de Salamanca, Spain
Jeroen de Bruin	Medical University of Vienna, Austria
Antonio Dourado	University of Coimbra, Portugal
Geert Gins	KU Leuven, Belgien
Warwick Graco	ATO, Australia
Aleksandra Gruca	Silesian University of Technology, Poland
Osman Hegazy	Cairo University, Egypt
Gary F. Holness	Quantum Leap Innovations Inc., USA
Pedro Isaias	Universidade Aberta, Portugal
Piotr Jedrzejowicz	Gdynia Maritime University, Poland
Joanna Jedrzejowicz	Gdynia Maritime University, Poland
Martti Juhola	University of Tampere, Finland
Janusz Kacprzyk	Polish Academy of Sciences, Poland
Mehmed Kantardzic	University of Louisville, USA
Dirk Krechel	University of Applied Sciences RheinMain, Germany
Mineichi Kudo	Hokkaido University, Japan
David Manzano Macho	Ericsson Research Spain, Spain
Dunja Mladenic	Jozef Stefan Institute, Slovenia
Eduardo F. Morales	INAOE, Ciencias Computacionales, Mexico
Jerry Oglesby	SAS Institute Inc., USA
Stefania Montani	Università del Piemonte Orientale, Italy
Jerry Oglesby	SAS Institute Inc., USA
Wieslaw Paja	University of Information Technology and Management in Rzeszow, Poland
Eric Pauwels	CWI Utrecht, The Netherlands

Table of Contents

Social Media Mining

Data Mining in Industry

Data Mining in Logistics

Data Mining in System Biology

Theoretical Aspects of Data Mining

Aspects of Data Mining

Risk Factors and Identifiers for Alzheimer's Disease: A Data Mining Analysis

Gürdal Ertek, Bengi Tokdil, and İbrahim Günaydın

Sabanci University, Faculty of Engineering and Natural Sciences, Istanbul, Turkey

Abstract. The topic of this paper is the Alzheimer's Disease (AD), with the goal being the analysis of risk factors and identifying tests that can help diagnose AD. While there exists multiple studies that analyze the factors that can help diagnose or predict AD, this is the first study that considers only non-image data, while using a multitude of techniques from machine learning and data mining. The applied methods include classification tree analysis, cluster analysis, data visualization, and classification analysis. All the analysis, except classification analysis, resulted in insights that eventually lead to the construction of a risk table for AD. The study contributes to the literature not only with new insights, but also by demonstrating a framework for analysis of such data. The insights obtained in this study can be used by individuals and health professionals to assess possible risks, and take preventive measures.

1 Introduction

The topic of this paper is the Alzheimer's Disease (AD) and the analysis of risk factors and identifying tests that can help diagnose AD. AD, a type of dementia disease, involves the irreversible degeneration of the brain which gradually ends up with the complete brain failure.

According to a 2012 report of the World Health Organization (WHO), 35.6 million people throughout the world are suffering from dementia diseases (Alzheimer Canada, 2013). Moreover, WHO projects that the total population of sufferers will double by 2030 and triple by 2050. It is also crucial to mention that, AD is the most common type of dementia disease. According to the statistics of Alzheimer's Association, AD accounts for 60 to 80 percent of the dementia cases (Alzheimer.org, 2013).

Neurodegeneration, progressive loss of neurons, increases due to aging and other factors, and these factors can lead to AD. On the other hand, neurodegenerative diseases such as AD cannot be diagnosed and treated fully due to the lack of treatment methods (Unay et al, 2010).

Besides the current statistics and forecasted spread of AD, the lack of a proven treatment method is another significant fact about this disease. Especially after the age of 65, AD generates a high risk to the population. A great percentage of the population suffers from this cureless disease, which eventually leads to death. Therefore, analysis of AD and insights based on available data are significant in understanding, alleviating the effects of, and paving the way to curing the disease.

P. Perner (Ed.): ICDM 2014, LNAI 8557, pp. 1–11, 2014.

Our study aims at generating a risk map of having AD after the age of 60. The probability of having AD will be analyzed in terms of age, social & economic status, gender, medical tests, and other factors, based on data coming from a field study. A detailed review of the literature on the factors that cause dementia and Alzheimer's disease can be found in a supplementary document (Supplement), and will not be included in this paper. Instead, we will focus on the work that we performed.

2   Data and Model

Our study uses data obtained from the Open Access Series of Imaging Studies (Marcus et al., 2010). The dataset consists of a collection of 354 observations for 142 subjects aged 60 to 96. Each patient may appear in more than one row. The subjects are all right-handed and include both men and women. The data also includes the education level and socio-economic status of the subjects. Moreover, some other medical statistics exist in the dataset, including intracranial volumes and brain volumes of the subjects. Summary statistics on the data, as well as some exploratory data analysis are presented in Marcus et al. (2010). We analyze the dataset using various visualization methodologies and a create risk map of the disease based on the given factors using classification trees.

Demented and *non-demented* are the classes in which the patient has the AD or not, respectively. *Converted* is the class that refers to the patients that develop the AD during the tests. The class *converted* was included in the classification tree analysis and cluster analysis, but removed from the dataset during the classification analysis. In classification tree and classification analyses, non-demented was selected as the target class, namely, the class that is predicted by the predictor attributes.

Table 1 presents the attributes (factors) in the analyzed dataset, explaining their meanings and providing their respective value ranges. Figure 1 presents the data mining process followed in the study. Figure 2 presents the roles assigned to attributes in the process ("Select Attributes" block of Figure 1). The attributes listed inside the "Attributes" box are the predictor attributes, whereas the attribute listed inside the "Class" box is the predicted attribute.

In Table 1, Clinical Dementia Rating is abbreviated "CDR". CDR can only take values 0, 0.5, 1 and 2. CDR being equal to 0 corresponds to non-demented subject CDR being equal to 0.5 corresponds to very mild dementia and CDR above 1 corresponds to moderate dementia. This medical test carries significance to entitle a subject as Alzheimer patient. The mini–mental state examination (MMSE) is a questionnaire test that has 30 questions. The goal of this test is to examine the cognitive situations of individuals. The questions of MMSE cover arithmetic, memory, and orientation. MR delay refers to the number of days between two medical visits. Other than those parameters, there is also information about age of the subjects in the classification tree. As mentioned earlier, the range of the age of the subjects is 60 to 96.

Classification tree analysis has been conducted with respect to the classes that the observed subjects belong to, namely demented, non-demented, and converted. In the data mining process (Figure 1), there are three main types of analysis: classification tree (decision tree) analysis, hierarchical clustering, and classification analysis. The data mining process begins with reading of the data from file (File block), and the validation of the data by displaying it in a data table, as well as observing the histogram (Distributions block), scatter plot (Scatterplot block), and attribute statistics (Attribute Statistics block). Then, each of the attributes is specified either as the class attribute or one of the predictor attributes (Select Attributes block). The roles specified for the attributes are given Figure 2.

The class label is "Group" and the key attribute is "MRIID". The attributes under the Attributes list box are predicator / factors in the classification tree analysis and classification analysis. In the clustering analysis, the Available Attributes "Visit", "MR Delay" and "CDR" are also included. The classification tree algorithm used is C4.5 and the created classification tree is visualized as a graph (Classification Tree Graph block). The visualized classification trees are displayed in Figures 3 and 4.

Hierarchical clustering analysis begins with the calculation of the attribute distances and storing these distances in a matrix (Attribute Distance block). Then hierarchical clustering is carried out (Hierarchical Clustering block). The visualization of the clusters is displayed in Figure 5.

Table 1. The explanation and the value ranges of the attributes of the OASIS dataset

Attribute	Explanation and Value Range
Group	The class label. Demented, non-demented, or converted.
MRIID	The test ID. Unique for each row. 1 to 354.
SubjectID	The subject's ID. 1 to 142. A subject may be visiting more than once, so the number of rows (354) is larger than the number of subjects (142).
Visit	Visit of the subject. 1 to 5.
MRDelay	The delay of a subject since the last visit.
CDR	Clinical Dementia Rating. 0 = no dementia, 0.5 = very mild AD, 1 = mild AD, 2 = moderate AD. (Morris, 1993)
Gender	Male (M) or Female (F)
Age	The age of the subject at the time of observation
EDUC	Education level
SES	Socioeconomic status, which is assessed by the Hollingshead Index of Social Position. 1 (highest status) to 5 (lowest status). (Hollingshead, 1957)
MMSE	Mini-Mental State Examination value. 0 (worst value) to 30 (best value). (Folstein, Folstein, & McHugh, 1975)
eTIV	Estimated total intracranial volume (cm3) (Buckner et al., 2004)
nWBV	Normalized whole-brain volume, expressed as a percent of all voxels (Fotenos et al., 2005)
ASF	Atlas Scale Factor; volume scaling factor for brain size.

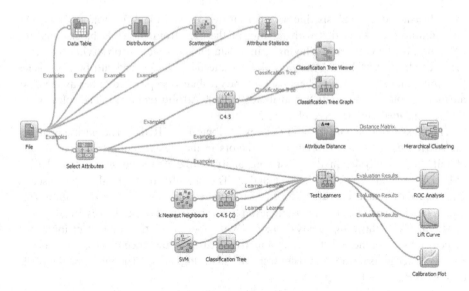

Fig. 1. The data mining process followed in the study

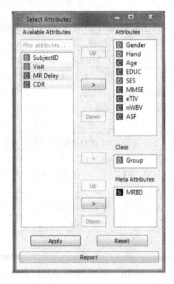

Fig. 2. The key attribute (Meta Attributes), class label (Class), and the attributes used for prediction (Attributes). The clustering analysis also includes the grey-shaded attributes within Available Attributes.

The classification analysis involves four classification algorithms (learners), namely k Nearest Neighbors, C4.5, SVM, and Classification tree. The performances of these four learners were compared (Test Learners block) with respect to classification accuracy, using a 5-fold design.

3   Analysis and Results

In this section, we present the data mining results and the insights that we obtain through these results. The analysis has been carried out using Orange (Orange) and Tableau (Tableau) data mining software. The analysis results are presented as a list of insights, and are later summarized in Table 2.

The preliminary classification tree constructed considered MRIID as the key attribute, Group as the class attribute, and included all the other attributes (except SubjectID) as factors. However, this resulted in a tree where the first split based on CDR perfectly distinguished the demented patients (CDR=1) from other subjects (non-demented and converted). This showed that CDR was too good of a factor to include in the analysis.

In the preliminary analysis, the next split in the tree was based on the attribute "MR Delay". However, using this attribute also had an inherent flaw: The demented subjects need to be under control with frequent medical tests. Most of the potential Alzheimer patients take the MR tests earlier than 675 days. Therefore "MR Delay" is dependent on the "CDR" score, and the probability of being converted. The subjects whose "CDR" values are greater than 0, and additionally if "MR delay" periods of these patients are smaller than 675 days, with 97.9% probability these subjects are either now or eventually became converted Alzheimer patients. The attribute "Visit" (number of visits) is also dependent on the "CDR" results.

Observing the "too perfect" results in the preliminary classification tree analysis due to "CDR" and the inherent dependency problem of "MR Delay" and "Visit", we decided to carry out our analysis by excluding these three attributes from the list of factors, as given in Figure 2.

Figures 3 and 4 show the graph visualizations of the classification tree after the attributes were selected as in Figure 2. In each pie, the light-colored slice represents the non-demented observations, darker slice represents demented observations and the darkest slice represents the converted observations (subjects who were observed not to have AD at that observation, but later possessed the disease).

In analyzing the classification tree graph, as visualized in Figures 3 and 4, we will be especially interested in two types of observations: 1) The deviations from the original distribution of the class labels (root of the tree), 2) The significant deviations between the parent and children nodes after a split is made.

The insights obtained from the classification tree analysis are now presented, following the observations that lead to those insights. Insights 1 through 6 are based on the expansion of the left mode (Figure 3), whereas insights 7 and 8 are based on the expansion of the right mode (Figure 4).

In Figure 3, the branches of the classification tree are split firstly (Split A) with respect to the values of MMSE. MMSE is thus a high-ranking indicator of AD. As it can be observed seen from the right branch of Figure 3, if MMSE<26, then the patient is demented at the time of the observation with a very high probability (94%). However, the left branch needs a further analysis.

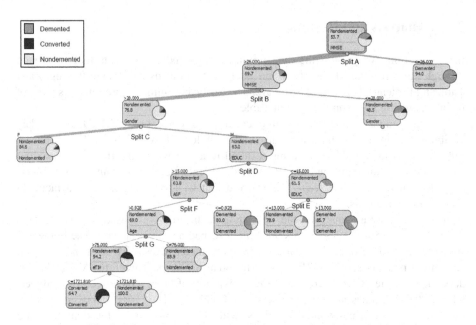

Fig. 3. The expansion of the classification tree for CDR >0.250

Insight 1: *If the MMSE value is smaller than 26, the risk of AD increases considerably to 94%.*

In Figure 3, in the left branch, the tree is first split based on MMSE again (Split B), and then based on gender (Split C). The MMSE values are greater than 28. In the female gender side of the branch, there is 84.6% probability of being non-demented.

Insight 2: *If a woman has MMSE value greater than 28, than she has a probability of being non-demented with a probability of 84.6% at the time of the observation.*

The branch of men is split further to make more analysis. The next split (Split D) is with respect to education values of the male people. Education level is divided into two. In the left branch, there are males with education level EDUC≤15 while in the right branch the education level EDUC≥15.

Insight 3: *Among the men who have MMSE>28, those who have an education level EDUC>15 have 63.8% chance of being non-demented at the time of observation, and those with EDUC≤15 have 61.5%. Yet there are no converted among those on the right branch; they are all demented. Therefore, less educated subjects show signs of dementia early on, whereas more educated convert later, summing to similar percentage of AD in the long run.*

Even though Insight 3 says that the percentage of demented plus converted is very close for Split D, Insight 4 goes into the detail, based on Split E.

Insight 4: *Among the men whose MMSE>28, those who have an education level in the range (13, 15] have much higher chance of having dementia, compared to those in other value ranges. Thus, the most risky range of education level for males who have MMSE>28 is the interval (13, 15], which refers to Bachelor's diploma at a university.*

When the branch of education level is EDUC>15 (left branch below Split D) is considered, there are again two other branches. These branches split according to their ASF values. ASF is the abbreviation of Atlas Scale Factor. This is a clinical term, which is the result of the MRI scans, and explained in Table 1.

Insight 5: *Among the men who have EDUC>15 and MMSE>28, those who have the ASF≤0.928, 80% are demented at the time of observation (right branch under Split F). Therefore, for men in this group, ASF is a major identifier of AD.*

The branch where the ASF value is greater than 0.928 is divided into two, according to Age (Split G). The left branch is where the age is greater than 76 while the right branch is the age is equal to or less than 76. In this study, the ages were between 60 and 96, and the age 76 seems to be the threshold age for men where significant changes take place.

Insight 6: *For men with EDUC>15, MMSE>28, and ASF>0.928, the age is equal to 76 or less than 76, there is 88.9% conditional probability that they are non-demented.*

This insight can also be expressed as follows: If a man with more than 15 years of education has MMSE>28 and ASF>0.928 when he is older than 76, then he will most probably *not* have AD.

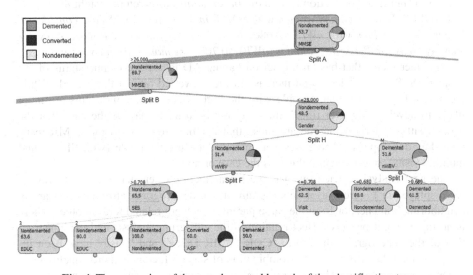

Fig. 4. The expansion of the non-demented branch of the classification tree

So far, the branch of the classification tree for the subjects who have MMSE value greater than 28 has been analyzed by observing Figure 3. Now the subjects who have MMSE value between 26 and 28 will be analyzed through Figure 4 (the right branch under Split B). This branch of the tree contains a greater portion of demented and converted subjects compared to the other branch. As the effect of the MMSE value has been indicated, the same effect can be observed in Figure 4. The smaller MMSE value results in higher risk of having the disease. The effect of other factors such as gender, SES and nWBV will be explored through Figure 4.

Gender could be an indicator for the statistical studies. When the non-demented percentages (under Split H) are compared with respect to gender, it can be seen that both genders have more or less the same percentage of non-demented subjects. Specifically, the female branch has 51.4% non-demented and male branch has 51.6% demented proportions. Therefore, there is not a clear distinction between these two values in terms of reasoning a differentiation. However, when the composition of the remaining portion of the pie is analyzed, it is observed that the remaining men (M) are almost all demented, whereas about half of the women (F) are converted later.

Insight 7: *For subjects that have MMSE in the range (26, 28], men and women exhibit similar percentages of non-demented, but almost all the remaining exhibit dementia at the time of observation, whereas nearly half of the women develop dementia later (being converted).*

The next important factor according to the classification tree graph is nWBV, which is an abbreviation for "Normalized whole brain volume". nWBV is the next splitting attribute for both men (M) and women (F), as can be seen in Splits I and F, respectively. For men, nWBV>0.680 signals a big risk factor, since 61.5% of the men under Split I, who have nWBV>0.680 are demented. For women, Split F tells that having nBWV≤0.708 completely guarantees being demented or being converted. There is a significant deduction from these observations, as given in Insight 8:

Insight 8: *When men and women with MMSE in the range (26, 28] are considered, larger nWBV values with nWBV>0.680 (larger brain volumes) are more risky for men, whereas smaller brain volumes (nWBV≤0.708) are more risky for women.*

The other factor that has an impact on having AD is socioeconomic status of the subjects, SES value. There is an increase in the converted ratio if the SES=1. While one might hypothesize that "People with the highest socioeconomic status are more likely to develop AD over time", this may not be true. It may be the case that the people with the largest income are those that continue to come to future MR tests, until they develop AD. There was not enough data in our sample (with SES=1 and multiple visits) to test whether this was the case or not.

The next analysis carried out was the hierarchical cluster analysis, whose results are displayed in Figure 5 as a dendrogram. In the dendrogram, attributes that are in neighboring branches, or from the same parent branch are related to each other. There is no input/output or cause/effect relation in the clustering analysis and the construction of the dendrogram; therefore, "CDR", "MR Delay" and "Visit" have been included among the attributes. The combination of several factors is more conclusive in terms of the risk map. The proximity of the attributes to each other can be seen through clustering. For instance, the education level of the subject is closely related with the socio-economic status of the subject (SES). In addition, both of SES and education are related to the result of mini-mental state examination (MMSE) of the subjects. Based on this observation from Figure 5, the relation of education to Alzheimer's deserves further investigation. Education directly influences the social status of the individuals in real life. Therefore, those two factors are connected to each other and the dataset of OASIS specifies these two data have similar effects. For further insights, the values for the education attribute "EDUC" can be discretized to take the following categorical values:

1: less than high school degree
2: high school degree
3: some college
4: college degree
5: beyond college

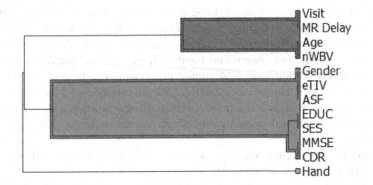

Fig. 5. Dendrogram for the attributes, showing their proximity to each other based on the sample data

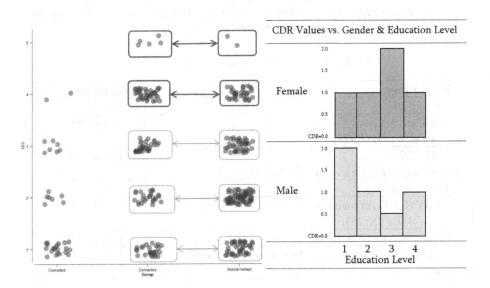

Fig. 6. Scatter plot of education effect

Fig. 7. The relation between gender, education, and CDR

In Figure 6, an analysis is performed to reveal the possible relation between the education levels and the risk of having Alzheimer. For this visualization, the density of points for non-demented and the density points for other values can be compared.

For education level taking values of 1, 2 or 3, the density of non-demented subjects is greater than the others, meaning that the risk of the disease is lower. However, for the education level values of 4 and 5, it can be seen that the number of demented subjects are greater than the non-demented subjects. For better illustration, the comparisons were highlighted in Figure 6. The insight obtained is the following:

Insight 9: *The risk of Alzheimer is higher for people with college degree or higher.*

Next, the effect of education level and the effect of gender were considered together, as shown in Figure 7. CDR is a crucial factor to identify AD. As it is evaluated before, if CDR is greater than 0.5, the possibility of having the disease increases considerably. The important observation for the Figure 7 is that the risk for women is greater for the education level 4. A similar observation can be done for the men for the education level 1. On the other hand, for the education level 4 the opposite observation can be made. Hence, it can be summarized that women with college degrees are in a riskier position than men with college degrees, in terms of being an Alzheimer patient.

Insight 10: *For higher education levels, especially for women college graduates are in a riskier position to having AD.*

The final analysis carried out was classification analysis, where the predictive power of the attributes has been tested. Unfortunately, the classification accuracies came out to be too low.

Insight 11: *The listed attributes cannot predict the risk of AD accurately.*

Therefore, as a deduction of Insight 11, one should rather focus on exploratory data mining for the given data, rather than predictive data mining.

Table 2. The summary of insights on the risk of AD

Risky ranges	Related Insight(s)
MMSE$\leq$26	Insight #1 & 2
EDUC$\in$ (13,15] *(for men with MMSE>28)*	Insight #3 & 4
ASF$\leq$0.928 *(for men with EDUC>15 and MMSE>28)*	Insight # 5
MMSE$\in$ (26,28]	Insight # 7
nWBV>0.680 for men (M), nWBV$\leq$0.708 for women (F) *(for MMSE$\in$ (26,28])*	Insight # 8
College degree or higher	Insight # 9
College degree for women, less than high school degree for men	Insight # 10

4   Conclusions

It is expected that the number of Alzheimer patients will increase in upcoming years. Apart from this projection, the lack of a precise medical treatment method for this disease will also continue to increase the possibility of deaths due to Alzheimer. Due to these facts, the understanding of AD risks is crucial.

As a contribution to the previous literature on AD, in this study, the effects of the factors are examined from a broader perspective through data visualization and mining methods. Rather analyzing brain images, the demographics and test statistics for the subjects have been examined. In terms of presenting the risk map of AD, the riskier ranges of each crucial factor can be summarized as in Table 2. As a distinctive factor from other studies of AD, our study is based on a recent dataset that includes not only demographic attributes, but also test results as attributes.

The study contributes to the literature not only with new insights, but also by demonstrating a framework for analysis of such data. Individuals and health professionals to assess possible risks, and take preventive measures can use the insights obtained in this study. The insights can also be used by health institutions, pharmaceutical companies, insurance companies, government institutions for planning their strategies for the current and the future.

Acknowledgements. The authors thank Precious Joy Balmaceda for her help in proofreading the paper.

References

1. Alzheimer Canada (2013), http://www.alzheimer.ca/en/Get-involved/Raise-your-voice/WHO-report-dementia-2012 (accessed on January 24, 2013)
2. Alzheimer.Org, http://www.alz.org/dementia/types-of-dementia.asp (accessed on January 24, 2013)
3. Marcus, D.S., Fotenos, A.F., Csernansky, J.G., Morris, J.C., Buckner, R.L.: Open access series of imaging studies: longitudinal MRI data in nondemented and demented older adults. Journal of Cognitive Neuroscience 22, 2677–2684 (2010)
4. Supplement for "Risk Factors and Identifiers for Alzheimer's Disease: A Data Mining Analysis", http://people.sabanciuniv.edu/ertekg/papers/supp/11.pdf
5. Unay, D., Chen, X., Erçil, A., Çetin, M., Jasinschi, R., van Buchem, M.A., Ekin, A.: Binary and nonbinary description of hypointensity for search and retrieval of brain MR images. In: IS&T/SPIE Electronic Imaging, Multimedia Content Access: Algorithms and Systems III, San Jose, California, USA (January 2009)
6. WHO, Dementia: a public health priority. World Health Organization and Alzheimer's Disease International (2012), http://www.who.int/mental_health/publications/dementia_report_2012/en/ (accessed on November 13, 2013)

Unsupervised Named Entity Recognition and Disambiguation:
An Application to Old French Journals

Yusra Mosallam[1], Alaa Abi-Haidar[2], and Jean-Gabriel Ganascia[2]

[1] DMKM Masters, UPMC Paris, France
[2] ACASA, LIP6, UPMC, Paris, France
alaa.abi-haidar@lip6.fr

Abstract. In this paper we introduce our method of Unsupervised Named Entity Recognition and Disambiguation (UNERD) that we test on a recently digitized unlabeled corpus of French journals comprising 260 issues from the 19th century. Our study focuses on detecting person, location, and organization names in text. Our original method uses a French entity knowledge base along with a statistical contextual disambiguation approach. We show that our method outperforms supervised approaches when trained on small amounts of annotated data, since manual data annotation is very expensive and time consuming, especially in foreign languages and specific domains.

1   Introduction

Named Entity Recognition (NER) was first introduced in Message Understanding conference 6 (MUC-6) in 1995 [1]. It aims at identifying the occurrences of general concepts including person, location and organization names or quantities like currencies, date and percentages in text. NER can also be viewed as a classification task that assigns each named entity (or entity mention) to its correct class, concept or entity. The identified named entities can be employed later in solving several problems such as: topic detection, automatic translation, automatic question-answering, semantic tagging, etc. NER techniques are usually tailored according to the following factors [2]:

- *Entity types* are also known as entity categories or concepts. The simplest NER techniques are used to identify coarse-grained concepts such as person, location and organization which are known as ENAMEX [1], which we study here. It is also common to identify fine-grained entities such as people of various professions, like doctors and professors, or specific people like Barack Obama [3] but that is beyond the scope of this paper.
- *Domains* such as business, medicine or information technology, and *text genre* like journalistic, scientific or informal genres have to be considered while designing a NER algorithm. Adapting NER techniques from one domain to another has bad implications on accuracy as shown in [4]. Various NER studies focus on news corpora, due to their abundance and public availability.

P. Perner (Ed.): ICDM 2014, LNAI 8557, pp. 12–23, 2014.
© Springer International Publishing Switzerland 2014

– *Language* is another important factor that needs to be considered. English has so far received the biggest attention in the NER domain with respect to other languages, however, masive amounts of texts in foreign languages are being published online [5]. French is not as popular as English and Older French text from the 19th Century is much more challenging due to the lack of annotated data and knowledge bases.

Several studies have been carried out to develop domain and language independent NER techniques, such as the one discussed in [6]. This method consists of 2 phases, the first is to use a naïve NER algorithm to tag concepts in the corpus. Then, it feeds the tagged corpus into the second phase which infers a contextual model for each concept using syntactic and semantic relations. Nevertheless, this method's accuracy depends heavily on the naïve NER algorithm used, it is noise prone, and it is not adequate in highly ambiguous context. Another bootstrapping language-independent NER algorithm was introduced in [5]. This iterative-learning algorithm uses a small list of unambiguous examples and unannotated training text to extract morphological and contextual clues. It was validated using 5 languages and it showed good precision but relatively low recall. Moreover, the algorithm heavily depends on the size of the training text.

According to [7], 16 language-independent NER systems were used to process English and German corpus in CoNLL-2003 Shared Task. Still, it depends heavily on annotated training corpus, and require a lot of manual work.

NER tasks can be classified according to the learning approach whether it is supervised or unsupervised. *Supervised* techniques depend on manually annotated data, and amongst the popular techniques are Hidden Markov Models (HMM), Conditional Random dom Fields (CRF), Maximum Entropy measures (ME), and Decision Trees. One of the major factors of success for such techniques is the availability of a huge annotated corpus in the target domain and language. However, manual annotation is very expensive in terms of human labour and time, and even sometimes inaccurate due to inter-annotator inconcistencies [8].

Unsupervised techniques can be divided into 2 categories; the first uses clustering, such as the one discussed in [9] which uses a generative model to cluster named entities. The second uses *Knowledge-based* resources to identify entities. Recently this approach became very popular due to the vast amount of knowledge sources available, and that it does not require manually-annotated corpus. Given our massive corpus of unlabeled data, we chose to investigate knowledge-based unsupervised NER techniques further. However, knowledge-based unsupervised NER techniques suffer from ambiguity if not from the limitation of knowledge bases that are finite and not continuously updated especially in the case of Old French Journals. Moreover, some knowledge bases are enriched with links between entities thus providing contextual information that could help resolve the ambiguity of matched entities. That is again not the case for currently available French knowledge bases that hardly match entities from the corpus of Old French Journals at hand.

Therefore, we use simple knowledge-base unsupervised NER techniques along with a statistical method to resolve ambiguity based on contextual information as discussed in the following sections.

The following sections are organized as follows: in section 2 we explore existing unsupervised NER techniques. In section 3 we explain our UNERD algorithm. In section 4 we describe the data, after which, our evaluation technique is explained in section 5. Preliminary results are presented in section 6, and finally we conclude and discuss future work in section 7.

2 Unsupervised NER

Knowledge-based techniques have recently attracted a lot of attention [10,11,12] thanks to the enormous amount of knowledge sources available. Using knowledge bases allows the identification of coarse-grained and fine-grained entity classes. However, the challenge remains in finding an appropriate knowledge base to use, and then to resolve ambiguity and infer the classes of rare entities. In this approach we divide the NER task into the following stages:

- *Preprocessing* focuses on preparing the corpus for the NER task and removing noise and unreliable data.
- *Entity Spotting* focuses on extracting all entity mentions from the corpus along with their contextual information to be used in the subsequent stages. We identify proper nouns using a pre-trained Part of Speech (PoS) Tagger. Other techniques use brute-force search with a variable size sliding window in order to indentify entity mentions, such as in [12].
- *Entity Matching* deals with finding all candidate matches of the entity mention in the knowledge base. In this stage we record three types of entities: (1) unique entities, which are matched with only one class (2) ambiguous entities, which are matched with more than one class, and (3) unmatched entities. The entity matching task usually involves a simple look-up algorithm using a knowledge base or a dictionary (also known as gazetteer).
- *Entity Disambiguation* Is a crucial complement to the matching step. It uses contextual information to disambiguate ambiguous entities and identify unmatched entities which, in our case, encompass more than 50% of the entity mentions. Hence, The overall performance of the NER technique depends heavily on the accuracy of the disambiguation step. We shall examine different disambiguation techniques and elaborate on our disambiguation approach in the following sections.

2.1 Knowledge Bases

Knowledge bases can be built automatically or manually with the help of communities as in the case of Freebase. The method discussed in [13] automatically generates gazetteers and uses them in identifying entities. It is based on the method described in [14] to extract named entities from web pages iteratively. It starts with 4 seeding examples of each category and searches for web pages that contain them. Then it uses a web page wrapper to extract more similar entities iteratively until it builds a sufficient knowledge base for each category. Other efforts were employed to automatically build entity knowledge bases from Wikipedia, Wordnet, and GeoNames such as YAGO and DBPedia. Aleda [15] is a French entity knowledge base extracted from the French

Wikipedia and Geonames. Selection of a knowledge base for NER is very important, and it depends heavily on the domain and language of the corpus. For our case we have a corpus of old French journals, hence, we select the Aleda knowledge base which was recently updated (17 April 2013).

2.2 Named Entity Disambiguation

Named Entity Disambiguation (NED) is a classification problem in which ambiguous entity mentions are disambiguated, often based on their context. Some knowledge bases provide contextual information with each entity which facilitate the disambiguation process. The method presented in [10] uses a Vector Space Model to represent the context of each ambiguous entity and resolves its ambiguity by comparing its contextual vectors with those of the matched DBPedia resources. Thus, the disambiguation problem is mapped into a ranking problem for which the importance of a given term is measured using term frequency, and Inverse Candidate Frequency that is adapted from Inverse Document Frequency.

Another popular NED technique is graph-based disambiguation which project the corpus into a graph $G(V, E)$ where vertices V represent the words in the corpus and the candidate entities, and edges E connect different words based on similarity and coherence measures. The technique presented in [12] builds a graph $G(E(d), R(D), sf)$, where $E(d)$ represents the set of candidate entities corresponding to all the surface forms discovered in document D; $R(D)$ is the set of directed edges between 2 linked entities extracted from Wikipedia; and sf is a scoring function based on a centrality measure. Another method [11] resolves ambiguity by maximizing a linear objective function that comprise: prior entity popularity, contextual similarities between a text mention and an entity, and overall coherence of the entities in the document. The authors of this study *ibid* projected this objective function into a graph problem; by constructing Entities-Mentions undirected weighted graph. This method was evaluated using a dataset of short news articles and it showed outstanding results.

The aforementioned techniques identify and disambiguate fine-grained entities using the contextual information provided in the knowledge base. However, Aleda does not record the relations between entities nor provides a sufficient contextual data with them. It only has a definition field which accompany some of the entities. Thus, adopting one of these techniques and relying on the limited contextual information in Aleda was infeasible.

According to [16], the similar approach of *word sense disambiguations* (WSD) can be targeted using machine learning techniques. WSD can be approached using bootstrapping semi-supervised techniques, such as the method presented in [17], which starts with a few seeding senses, and labels all their occurrences in the corpus automatically using a small seed of collocations (words that co-occur with the word sense/ class). Then it uses decision lists to label more examples, and extract new significant collocations. The algorithm proceeds iteratively until all word senses are labelled. WSD was also targeted using unsupervised techniques, such as contextual clustering at which each word is represented by a contextual vector and then a clustering algorithm is applied to these contextual vectors [16]. Other unsupervised techniques uses word clustering, by measuring semantic similarities between different words based on the syntactic dependencies in the corpus *ibid*.

Our use of unsupervised knowledge-base techniques is motivated by the recent digitization and publication of Old French journals resulting in huge, unlabelled and very expensive to annotate data. The French language used in the corpus in old and different from the contemporary French. Hence, we choose to use the Aleda knowledge base with statistical disambiguation techniques to identify ambiguous and unknown entities. More specifically, we adopt a variant of a bootstrapping WSD algorithm [17] by considering the collocations of uniquely identified entities. Our algorithm UNERD, which we present in the following section, yields competitive results.

3   UNERD Algorithm

As shown in figure 1, we begin with text preprocessing, which we describe in section 4; Then, we spot possible entity mentions in the corpus and use Aleda to find all possible matches for them. Next, we extract salient words from the uniquely matched entities and use them to resolve ambiguity and to identify unmatched entities.

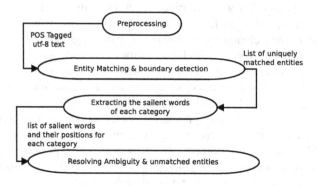

Fig. 1. Algorithm Overview

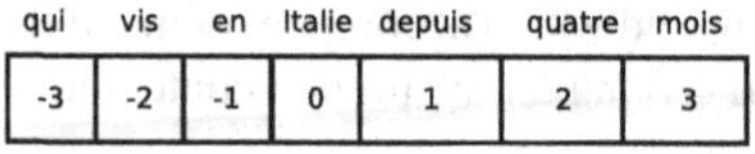

qui	vis	en	Italie	depuis	quatre	mois
-3	-2	-1	0	1	2	3

Fig. 2. Context of a mention of "Italie"

3.1   Matching and Boundary Detection

In the matching phase we process the PoS tagger's output to find all candidate entities as follows. We start by considering each word within a noun phrase separately and finding the possible matching entities for it. After that, we assign to each noun phrase a single entity type, if possible, based on the tags assigned to its contents. Before we assign a tag to a noun phrase we search Aleda for a possible matching entity for the whole noun phrase. Then, if an exact match is found we assign it to the noun phrase, else we assign the infered tag to it. Hence, we implicitly solve the problem of ambiguity and unmatched entities within noun phrases.

Some of the French named entities are composed of multiple noun phrases connected by a proposition such as "avenue des Champs-Elysées". Thus, after matching single noun phrases we examine each two consecutive noun phrases connected by a proposition and merge them into a single entity according to the three following conditions: 1) The two noun phrases belong to the same category. 2) One of the noun phrases belongs to a unique category X , whereas the other one does not belong to a category but contains a frequent word (extracted from Aleda) of category X. 3) The first noun phrase is tagged as a person (category), while the second is tagged as a location (category), for example "Jean de Sainte-Colombe", in which case we merge them as a person's name.

Before tagging the merged named entity we search for it in Aleda for possible exact matches. If an exact match is found it is given preference over the inferred tag.

3.2 Disambiguation

The use of Aleda knowledge base limits our choices of disambiguation techniques as it does not provide relations between entities nor contextual evidence for entities or categories. Therefore, we use cues from uniquely matched entities in text to resolve ambiguity of ambiguous entities and identify unresolved entities. We consider the words occurring before and after uniquely matched entities, as shown in figure 2. We exclude all the punctuations and non-alphabetical strings from extracted words. Then, we calculate discriminative words for each category for each contextual-position using the following scoring methods:

- *TF.IDF* is an overall measure of the discriminative power of a term. $TF.IDF = TF_a(w_j) * IDF_a(w_j)$ where $TF_a(w_j) = 0.5 + 0.5 * \frac{tf}{max(tf)}$ is the augmented term frequency [18], $tf_a(w_j) = \frac{n_j}{N_a}$ that is the classical term frequency, and $IDF_a(w_j) = log\frac{N}{n_a}$ is the Inverse Document Frequency (IDF).
- *Relative Term Frequency (RTF)* is another discriminative factor that gives the frequency of a term w_j occurring in a class a relative to its occurrences in other classes giving a less restricted term discrimination measure than the IDF, as follows $RTF(w_j) = \frac{TF_a(w_j)}{\sum_{b \in C, b \neq a} TF_b(w_j)}$ where C is the set of possible classes. We uses RTF to compensate for the OCR errors and imprecisions in the knowledge base matching.

We use both the RTF and TF.IDF to select a list of salient words for each category, which can help in resolving ambiguity and identifying unmatched entities. Salient words selection is based on the following approaches:

- *Bag of Words* picks the class that has the highest TF-IDF ignoring the relative positions of the contextual words with respect to the entity-mention.
- *Single Significant Cue* considers each location separately and gets the category with the highest TF.IDF for each position. Then it select the position with the highest TF.IDF to determine the category.

– *Majority Vote* The same as the previous approach, except that it selects the category with highest position-votes.

Finally, we resolve the ambiguity by combining the outcome of the 3 aforementioned approaches if at least 2 of the 3 approaches agree on the same decision.

4   Data

The corpus consists of old unlabelled French journals, more specifically, a subset of "Le Petit Parisien" a journal supplied by Bibliothéque nationale de France (BnF). It was published between 1863 and 1944 with a total of 29616 issues. The corpus we are using comprises 260 issues with a total of 1098 pages of natural French text. The pages are encoded in ALTO format, which is an open XML standard for representing OCR text.

The ALTO format provides OCR confidence for each word. The XML files are encoded using the iso-8859-1 encoding. Handling this dataset is challenging due to the old manuscripts and the resulting OCR errors. However, we exclude text blocks with low OCR confidence (lower than 0.85).

The corpus of French journals is very expensive and time consuming to annotate. Therefore we had experts annotate only a small portion of 4171 words of which the numbers annotated entities are listed in table 1

Data preprocessing The first essential step is to prepare OCR text for entity extraction such as the XML iso-8859-1 encoded corpus into standard UTF-8 encoded text, excluding text-blocks with confidence less than 85% or containing words with length more than 15 characters, applying Part of Speech Tagging (using TreeTagger [19] and removing French stop-words.

5   Evaluation

5.1   Metrics

NER Classification performance is usually evaluated by comparing its outcome to a corpus annotated by human linguists or experts. In MUC there are 2 main measures for evaluating NER techniques, as mentioned in [2]. The first one is the correct assignment of class to the entity regardless of its boundaries. The second measure is the correct detection of entity's boundaries regardless of the type. Here, we use the first measure and compute precision and recall. The overall evaluation of the system is calculated using f-measure (or F-score) which is the harmonic mean of the precision and recall, namely, $F_1 = 2 * \frac{precision * recall}{precision + recall}$ For more information about the F-measure or alternative evaluation and scoring techniques please refer to [20].

5.2   Cross-Validation

We use the traditional K-fold cross-validation in which the annotated data is divided in to K parts: K-1 for training and 1 for validation in order to test for how well our model predicts and generalises on separate data. More specifically, we use a 4-fold cross-validation approach in which we split the annotated trained data in 4 parts or folds. Our method is trained on 3 out of 4 folds (i.e. 75%) of the annotated data and tested or validated on the remaining fold (i.e. 25 %) of the annotated data. This is done repeatedly for all possible combinations in order to obtain several classification results [20]. The cross-validation is performed on a small section of Manually Annotated Data (MAD)[1] that we had experts annotate manually. Finally, the results are computed and averaged using traditional metrics as discussed previously.

5.3   Alternative Methods

We explore four different scenarios, namely, S1, S2, S3, and S4 along with our UNERD method. These scenarios describe several supervised and unsupervised methods (combined and varied) in order to to compare our UNERD method and understand its different behaviors. The 5 scenarios are illustrated in figure 3 and explained as follows:

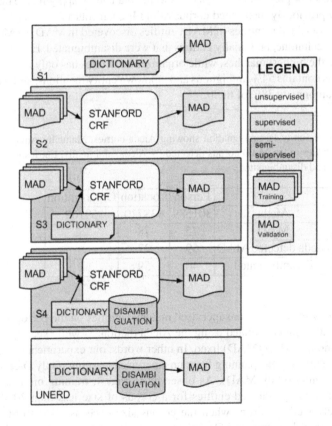

Fig. 3. Supervised, unsupervised, and semi-supervised scenarios tested and compared

---

[1] The MAD consists of only 4171 words of which the numbers if matched entities are listed in table 1.

- Scenario (S1) is the baseline method, where we only use the dictionary to tag entity mentions.
- Scenario (S2) simply uses 75% of the Manually Annotated Data (MAD) for CRF classification training and tests on the remaining annotated data.
- Scenario (S3) is like S2 but in addition to training on a portion of the MAD it uses a dictionary to tag additional data for training.
- Scenario (S4) is similar to S3, however it adds to the dictionary look-up of S3 our UNERD in order to disambiguate the dictionary tagged text and make it more reliable for training.
- Finally, we present our unsupervised method of UNERD that uses no training and that is discussed previously.

6   Results

In this section we show preliminary results that were obtained by applying the proposed technique to the previously mentioned corpus of old French Journals. Table 1 lists the number of entities in Aleda, entities in MAD, entities discovered in MAD by Aleda and by our matching technique, and finally entities that were disambiguated. Evidently, person entities have the highest matches, while organization entities has only few matches. The MAD corpus contains 54% ambiguous and unknown entity mentions, and we could successfully disambiguate 78% of them.

Table 1. Statistics of Entity Disambiguation showing Aleda entries, Manually Annotated Data Entries (MAD), Entries in the MAD corpus detected by Aleda, and finally, Entries in the MAD detected by Aleda and disambiguated

	Person	Location	Organization
Aleda	301589	438710	59169
MAD	75	78	22
Aleda detected MAD	69	27	9
Disambiguated	40	59	2

In order to study the effect of unsupervised matched entities on the performance we varied the amount of data annotated using the Aleda dictionary while keeping the size of Manually Annotated Data (MAD) fixed. In other words, our experiment can be seen as a one of semi-supervision spanning from unsupervised learning (Only Dictionaries) to supervised learning (Only MAD). As observed in Fig. 4, training on a fixed size of MAD with dictionary-matched entities for a corpus of size less than 2000 words perfoms relatively well but drops when the corpus size exceeds 4000 words for all three classes: Person, Location and Organization. However, S4 in Fig. 5, which adds disambiguation to the dictionary matching, shows a more stable performance over a larger corpus. That is evident in all categories except for the Organization category which is statistically insignificant due to few occurrences of Organization entities as

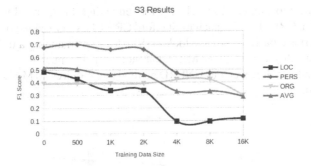

Fig. 4. S3 algorithm performance over a varying range of dictionary-tagged training data

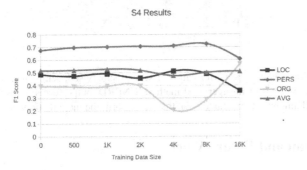

Fig. 5. S4 algorithm performance over a varying range of dictionary-matched & disambiguated training data

shown in table 1. The results are intuitive since many ambiguous dictionary matched entities are highly likely to negatively bias the learned rules from the MAD.

As previously mentioned, dictionary lookup is very basic and suffers from a finite number of entries that could have multiple senses and require disambiguation. As illustrated in Fig. 6, S1, representing mere dictionary lookup performs the worse for all 3 categories. Supervised machine learning is capable of learning rules from an anno-

Table 2. Evaluation results illustrated in Fig. 6. S1 is the dictionary lookup. S3 is trained on MAD and a 5k dictionary-annotated corpus with no disambiguation whereas S4 (1K) is trained on MAD and a 1k dictionary-annotated corpus with disambiguation. Finally, UNERD (8) does not train on MAD but uses an window of size 8-words (4 on each side) for disambiguation. A wide range of window sizes was tested, however, the performance results did not vary much.

	S1	S2	S3 (5k)	S4 (1K)	UNERD (8)
LOC	0.33	0.54	0.46	0.55	0.77
PERS	0.73	0.75	0.75	0.75	0.83
ORG*	0.20	0.44	0.44	0.43	0.46
AVG	0.42	0.58	0.55	0.58	0.69

tated text in order to predict entity classes for remaining unannotated text. As illustrated in Fig. 6, a CRF classifier such as Stanford CRF Classifier trained on a portion of MAD outperforms a basic dictionary-lookup (S1). Moreover, adding a training corpus of dictionary-matched entries to the MAD does not influence much the performance as shown for S3 and S4. However, our UNERD method that requires no MAD outperforms the remaining methods.

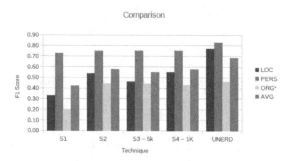

Fig. 6. Comparision between the performance of several supervised, semi-supervised and unsupervised methods, where our unsupervised method, UNERD, outperforms S2, S3 and S4 when trained on a small amount of annotated data, that outperform the mere dictionary-lookup (S1)

7   Conclusion and Future Work

The unsupervised use of dictionary-lookup is known to enhance NER, however dictionaries have limitations for being finite and ambiguous. On the other hand, supervised NER such as Stanford's NER Classifier that we tested here is known to perform very well but only with the availability of huge amounts of manually annotated training data that is very costly, time consuming and sometimes inaccurate due to inter-annotator inconcistencies. Therefore, we developed and discussed an original unsupervised approach for Named Entitiy Recognition (NER) and Disambiguation (UNERD) using a French knowledge-base and a statistical contextual disambiguation technique that slightly outperformed Stanford's NER Classifier (when trained on a small portion of manually annotated data) and Aleda's dictionary lookup. Arguably, a larger set of training data would enhance the results obtained by Stanford, however, MAD are costly and not always available, especially for Old French Text. Our NER and disambiguation technique is supposed to work on various languages and domains using unlabled data.

References

1. Grishman, R., Sundheim, B.: Message understanding conference-6: A brief history. In: COLING, vol. 96, pp. 466–471 (1996)
2. Nadeau, D., Sekine, S.: A survey of named entity recognition and classification. Lingvisticae Investigationes 30(1), 3–26 (2007)
3. Fleischman, M., Hovy, E.: Fine grained classification of named entities. In: Proceedings of the 19th International Conference on Computational Linguistics, vol. 1, pp. 1–7. Association for Computational Linguistics (2002)

4. Poibeau, T., Kosseim, L.: Proper name extraction from non-journalistic texts. Language and Computers 37(1), 144–157 (2001)
5. Cucerzan, S., Yarowsky, D.: Language independent named entity recognition combining morphological and contextual evidence. In: Proceedings of the 1999 Joint SIGDAT Conference on EMNLP and VLC, pp. 90–99 (1999)
6. Cucchiarelli, A., Velardi, P.: Unsupervised named entity recognition using syntactic and semantic contextual evidence. Computational Linguistics 27(1), 123–131 (2001)
7. Tjong Kim Sang, E.F., De Meulder, F.: Introduction to the conll-2003 shared task: Language-independent named entity recognition. In: 7th Conference on Natural Language Learning at HLT-NAACL, pp. 142–147. Association for Computational Linguistics (2003)
8. Bayerl, P.S., Paul, K.I.: Identifying sources of disagreement: Generalizability theory in manual annotation studies. Computational Linguistics 33(1), 3–8 (2007)
9. Elsner, M., Charniak, E., Johnson, M.: Structured generative models for unsupervised named-entity clustering. In: Proceedings of Human Language Technologies: The 2009 Annual Conference of the N. American Chapter of the Association for Computational Linguistics, pp. 164–172. Association for Computational Linguistics (2009)
10. Mendes, P.N., Jakob, M., García-Silva, A., Bizer, C.: Dbpedia spotlight: shedding light on the web of documents. In: Proceedings of the 7th International Conference on Semantic Systems, pp. 1–8. ACM (2011)
11. Hoffart, J., Yosef, M.A., Bordino, I., Fürstenau, H., Pinkal, Taneva, G.: Robust disambiguation of named entities in text. In: Proceedings of the Conference on Empirical Methods in Natural Language Processing, pp. 782–792. Association for Computational Linguistics (2011)
12. Hakimov, S., Oto, S.A., Dogdu, E.: Named entity recognition and disambiguation using linked data and graph-based centrality scoring. In: Proceedings of the 4th International Workshop on Semantic Web Information Management, p. 4. ACM (2012)
13. Nadeau, D., Turney, P.D., Matwin, S.: Unsupervised named-entity recognition: Generating gazetteers and resolving ambiguity. In: Lamontagne, L., Marchand, M. (eds.) Canadian AI 2006. LNCS (LNAI), vol. 4013, pp. 266–277. Springer, Heidelberg (2006)
14. Cohen, W.W., Fan, W.: Learning page-independent heuristics for extracting data from web pages. In: AAAI Spring Symposium on Intelligent Agents in Cyberspace (1999)
15. Sagot, B., Stern, R., et al.: Aleda, a free large-scale entity database for French. In: Proceedings of LREC 2012 (2012)
16. Navigli, R.: Word sense disambiguation: A survey. ACM Comput. Surv. 41(2), 10:1–10:69 (2009)
17. Yarowsky, D.: Unsupervised word sense disambiguation rivaling supervised methods. In: Proceedings of the 33rd Annual Meeting on Association for Computational Linguistics, pp. 189–196. Association for Computational Linguistics (1995)
18. Manning, C.D., Raghavan, P., Schütze, H.: Introduction to information retrieval, vol. 1. Cambridge University Press (2008)
19. Schmid, H.: Improvements in part-of-speech tagging with an application to German. In: Proceedings of the ACL SIGDAT-Workshop (1995)
20. Feldman, R., Sanger, J.: The text mining handbook: advanced approaches in analyzing unstructured data. Cambridge University Press (2007)

Multiple Template Detection Based on Segments

Bo Gao* and Qifeng Fan

Shenzhen Key Lab for Cloud Computing Technology and Applications
Peking University Shenzhen Graduate School, Shenzhen, Guangdong, P.R. China
{pkubobo1991,fanqf1026}@gmail.com

Abstract. Web pages contain a combination of informative contents and redundant contents which are primarily used for navigation, advertisements, copyright and decoration. Detecting templates correctly and precisely thus becomes a vital part for many applications. Methods for template detection have been studied extensively. However, they are insufficient to detect multiple templates in a Web site. In this paper, we propose a novel segment-based template detection method to identify templates. Our method works in three steps. First, for each Web site we construct a SSOM (Site-oriented Segment Object Model) tree from sampled pages in a Web collection, through aligning the pages' SOM (Segment Object Model) trees. Second, we construct a template from the SSOM tree. At last, the template can be used to detect templates for the Web site: Given a page in the Web site, its template contents are gained with mapping between its SOM tree and the SSOM tree and classifying. The proposed method is evaluated with two mining tasks, Web page clustering and classification. It leads to a significant improvement when compared to previous template detection methods.

Keywords: Template Detection, Information Extraction, Segmentation.

1 Introduction

The World Wide Web has long become a huge container of information, which includes news and reports about politics, economics, culture, entertainment and others. Recently, developments of forum have further greatly increased the magnitude of information. In the face of the large amount of information, many websites, especially commercial websites, provide web pages with much template contents for many purposes, such as banner advertisements, navigation bars, copyright notices, decoration, etc. Gibson et al. [7] have found that 40-50% of the content on the Web is template content and the volume is still growing steadily. Although templates are helpful for users to browse the Web pages, they are harmful to many web mining and searching systems. They can decrease the precision of search, increase the size of index and impair the performance of applications that manipulate web pages. If the templates can be removed, many applications such as link analysis and content extraction can gain a higher

---

* Corresponding author.

P. Perner (Ed.): ICDM 2014, LNAI 8557, pp. 24–38, 2014.
© Springer International Publishing Switzerland 2014

precision. And many applications can realize a significant improvement in performance. Thus it is very important to identify templates correctly and efficiently.

In this work, we focus on discovering informative contents based on the following observation: In a given Web site, templates usually share some common presentation styles. Moreover, the contents of templates tend to be similar or almost identical.

Many previous extraction methods we found in literature extract informative contents of Web pages based on per Web page analysis. These "page-level" template detection methods have some drawbacks. Generally, "page-level" methods have lower accuracy than "site-level" methods. Moreover, some heuristics page-level methods rely on manual rules and extract template effectively only in specified sites. Our method is based on a site-oriented structure which is the alignment of DOM trees of pages within a site. Template detection methods based on RTDM (Restricted Top-Down Mapping) are proposed in [11] [13]. They are motivated by the heuristics that: for each two input pages, if a sub tree that spans from the document root is detected in both pages, then it is regarded as a template. While these methods are efficient, they are of limited use because of the following two reasons: First, in some Web sites, especially article-type sites, many informative contents have almost identical structures, and they tend to be detected as templates in these methods. Second, since these methods take no text contents of Web pages into consideration, they cannot utilize them to distinguish informative contents from template contents. Yi et al. [16] and Li and Liu [15] propose a template detection method based on SST (Site Style Tree) by aligning the DOM (Document Object Model) trees to extract structural information of Web sites. If a node in SST is an internal node, its importance is given by combining its presentation importance and its descendants' importance. For a leaf node, its importance is given by its content importance. However, the method is unable to detect the informative contents in the multiple-templates Web sites very well, because the document collection of one SST node may be the combination of templates and informative contents.

We observe that generally templates appear in the form of segments. For example, segments for navigation, copyright and advertisements appear as template, and segments for main article appear as informative contents. In this paper, we use the segment as the granularity of template detection.

In addition, another contribution of this paper is that our method can effectively detect multiple templates in the same site through a machine learning algorithm.

The rest of this paper is organized as follows. In section 2, we give an overview of the problem of informative contents extraction and review related work. Then we illustrate site-oriented segment object model in Section 3. In section 4, we describe the approach for template detection. And experiments are discussed in Section 5. Finally, we draw a conclusion of this study and outline directions in Section 6.

2   Related Work

The problem of detecting templates and extracting informative contents of Web pages has been addressed in many previous research papers. Most of the proposed methods are based on machine learning or heuristics.

The problem was first discussed by Bar-Yossef and Rajagopalan in [1], proposing two template detection algorithms based on DOM tree segmentation and segment selection. While their proposal was considered as a segmentation method, the goal was to detect noisy information of Web pages. Their heuristics used the cues provided by hyperlinks. If an element contains more than K links, it is considered as a segment, otherwise it is counted as part of a segment containing it. And then template segments were selected by one of the above two template detection algorithms. A similar method was proposed by Ma et al. in [10], which segments pages by *table text chunk*. All table text chunks are identified as template table text chunk if their document frequency is over a determined threshold.

Wang et al. [14] proposed DSE (Data-rich Subtree Extraction) algorithm to recognize and extract the informative contents of Web page by matching simplified DOM trees. Lin and Ho [9] proposed a method to discover informative contents based on the heuristics that redundant blocks, opposite to informative blocks, appear more frequently. Reis et al. [11] presented a method based on the RTDM algorithm through generating patterns to identify templates. A similar solution was presented by Vieira et al. in [13], which detects templates through the operation of each step on the Tree Edit Distance of page structures.

Some other methods using the DOM tree of pages to extract informative contents were proposed in articles [16] [15] [7].

Yi et al. [16] and Li and Liu [15] studied noisy information elimination by aligning the DOM trees to extract structural information of Web sites. They present a data structure called SST (Site Style Tree) to capture the actual contents and the common presentation styles of the Web pages in a Web site. After the SST is constructed, the importance of each node is evaluated through information theory (or entropy). For each node in the SST tree, the more diversity of presentation styles and contents associated to it, the more likely the node is informative. On the contrary, less diversity indicates the node is likely to be noisy.

Gibson et al. [7] have conducted an extensive survey on the use of templates on the Web which revealed the rapid development of template. They also develop new randomized algorithms (DOM-based algorithm and Text-based algorithm) for template extraction. In DOM-based algorithm, for each node, the hash is computed by the content of the node and the start and end of offsets. And then, the nodes are considered as templates if the occurrence counts of their hashes are within a specified threshold. In Text-based algorithm, the page is pre-processed to remove all HTML tags, comments, and text within <script> tags, and then a shingling procedure is performed on the result data through sliding a window of size W. Templates are identified by comparing the occurrence count of the fragment contained in the window to a specified threshold.

Some page-level algorithms of template detection have been proposed in [4] [5] [8] [12].

In 2005, Debnath et al. [5] proposed a page-level algorithm ("L-Extractor") that uses various block-features and trains a Support Vector (SV) based classifier to identify an informative block versus a non-informative block. Kao et al. [8] proposed an algorithm to utilize entropy-based Link Analysis on Mining Web Informative Structures. Song et al. [12] used VIPS (Vision-based Page Segmentation) algorithm to segment a Web page into blocks which are then judged on their salience and quality. Chakrabarti et al. [4] developed a framework for the page-level template detection problem which is built on two main ideas: the automatic generation of training data for a classifier that assigns templateness score to every DOM node of a page, and the global smoothing of per-node classifier scores.

Many authors have explored Web page segmentation. While segmentation of pages and content extraction may appear different, there are many similarities between these two tasks, such as models and algorithms. And we can also extract informative contents through segmentation. One of the best solutions found in literature, called Vision-based Page Segmentation algorithm (VIPS), was presented by Cai et al. in [3]. The method segments a page by simulating the visual perceptions of the users about that page. But it requires a considerable human effort, since it is necessary to manually set a number of parameters to each page to be segmented. In [6] David addressed a method to segment Web pages of a site by aligning the DOM trees, which gives us inspiration.

We here propose a novel method based on SSOM for automatically detecting templates. This method leads to a significant improvement, compared to DOM Based Algorithm in [7], Text Based Algorithm in [7], SST Based Algorithm in [16] [15] and RTDM Based Algorithm in [13]. Our experiments indicate that our method can outperform these four baselines at F score.

3 Site-oriented Segment Object Model

Web pages usually contain informative contents and redundant contents. In a Web site, redundant contents tend to appear in many pages having the similar structure, text content and presentation style. In this section, we describe the site-oriented segment object model.

3.1 Representation of Web Pages

Unlike plain text document, Web pages consist of text contents and tags. Each page can be represented as a DOM (Document Object Model) tree, in which tags are internal nodes and the detailed texts, images or hyperlinks are leaf nodes. In this paper we expand the original DOM tree by adding some additional attributes to the nodes. We call the modified structure as SOM (Segment Object Model) tree, the node of which is defined as follows:

Definition 1 (SOM Node): A SOM node N has six components, denoted by $(tagName, content, styleHash, label, parent, children)$, where

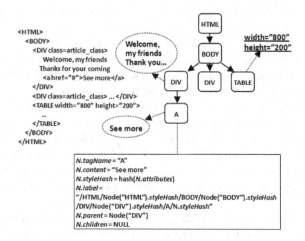

Fig. 1. An Example of SOM Tree

- *tagName* is the tag name, e.g., "DIV" and "TABLE";
- *content* is the text content of the node;
- *styleHash* is the hash of its attributes;
- *label* is the string generated from its node path and *styleHash*, e.g.,
/HTML/Node("HTML").*styleHash*/BODY/Node("BODY").*styleHash*/DIV
/Node("DIV").*styleHash*/A/Node("A").*styleHash*
(Detailed description will be given below.)
- *parent* is the pointer to its parent;
- *children* is the list of pointers to its children.

Figure 1 shows the HTML source code of a Web page and its corresponding DOM tree. In the figure, the circle is the actual content of the node. For example, for the tag "DIV", the actual contents are "Welcome, my friends" and "Thanks for you coming"; for the tag "A", the actual content is "See more". Underlined text is the display style of the node, e.g., for the tag "TABLE", its style is represented by attributes "width" and "height". Each node in DOM tree contains information about the name of the tag (*N.tagName*), and we use this information to recursively define the label of each node (*N.label*) as the concatenation of the label of its parent and its tag (*N.tagName*). Whenever two sibling nodes get equal *tagName*, we distinguish them by adding the *styleHash* to their label values. Therefore, for each two Sibling nodes, if they have the same *tagName* and *styleHash*, they have the same *label*, or else have different *labels*.

3.2   Representation of Web Sites

A SOM tree is sufficient for representing the content and style of a single HTML page, but it is insufficient to study the overall presentation style and contents of a set of HTML pages.

To address this problem, we create a hierarchical structure named SSOM tree (SSOM is an acronym to Site-oriented Segment Object Model) that summarizes the SOM trees of all pages found in a Web site, which is similar with SOM David proposed in [6].

Definition 2 (SSOM Node): A SSOM node S in SSOM tree is the combination of page nodes having the identical *label*; it has eight components, denoted by (*tagName, content, styleHash, label, parent, children, counter, classifier*), where

- *tagName* is the tagName of a page node;
- *content* is the set of content of SOM nodes containing it;
- *styleHash* is the styleHash of a page node;
- *label* is the label of a page node;
- *parent* is the pointer to its parent;
- *children* is the set of pointers to its children;
- *counter* is the number of pages containing it;
- *classifier* is the 0-1 classifier of S, which can be used to classify a segment

into *template* or *informative*.

4 Approach for Template Detection

Figure 2 shows modules of our approach. Each module will be described in the following sub-sections.

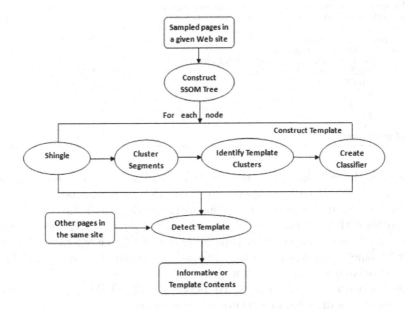

Fig. 2. The Modules of Template Detection Approach

4.1 Constructing SSOM Tree

The first step of our algorithm is constructing the SSOM tree. The *Insert* and *Merge* algorithm is described in Algorithm 1 and Algorithm 2. In Algorithm 3, we describe the *Construct* algorithm.

Algorithm 1. Insert

Input:
- The parent of the SSOM node which is prepared to be created: S_{parent}
- The SOM node which is prepared to be inserted: N

1: **procedure** :
2:     Create a new SSOM node S.
3:     $S.tagName \leftarrow N.tagName$
4:     $S.label \leftarrow N.label$
5:     $S.styleHash \leftarrow N.styleHash$
6:     $S.counter \leftarrow 1$
7:     Add $N.content$ to $S.content$.
8:     **if** S_{parent} is not null **then**
9:         Add S to $S_{parent}.children$.
10:     **end if**
11:     **for** each child n in N **do**
12:         Insert(S, n);
13:     **end for**
14: **end procedure**

Algorithm 2. Merge

Input:
- The SSOM node which N is prepared to be merged into: S
- The SOM node which is prepared to be merged: N

1: **procedure** :
2:     $S.counter \leftarrow S.counter + 1$
3:     **for** each child n in N **do**
4:         **if** $\exists s \epsilon S.children$ and $n.label=s.label$ **then**
5:             Add $n.content$ to $s.content$.
6:             Merge(s, n);
7:         **else**
8:             Insert(S, n);
9:         **end if**
10:     **end for**
11: **end procedure**

With the above 3 algorithms, the SSOM tree of the site can be easily gained. Based on the SSOM tree, we apply a segmentation method proposed in [6] to segment Web page. The segmentation algorithm works as follows:

• Dealing with nested content: Whenever we find internal nodes with textual content, we do not represent their children in our DOM tree representation, and thus it becomes a leaf node. The textual content associated to the removed tags is then associated to this new leaf node.

• Dealing with recurrent regions: We perform in the DOM trees is to remove sequences of tags disposed in a recurrent (or regular) way.

Thus, every leaf node of the SSOM tree is a collection of segments. And every internal node can also be considered as a collection of segments too.

We omit the detailed description due to the limitation of the paper length.

Algorithm 3. Construct

Input:
- Sampled pages in a given Web site: C

Output:
- The root of SSOM tree: S

1: **procedure** :
2:     Create an empty SSOM tree S_t, and set S is the root of S_t.
3:     **for** each page β in C **do**
4:         $N \leftarrow$ the root node of SOM tree of β
5:         Merge(S, N);
6:     **end for**
7:     Return S;
8: **end procedure**

Algorithm 4. ConstructTemplate

Input:
- The SSOM tree

1: **procedure** :
2:     **for** each node S in the SSOM tree **do**
3:         {Phase 1: Shingling}
4:         **for** each element e in $S.content$ **do**
5:             $shingle\_list \leftarrow shingle\_list \cup$ shingle (W, e)
6:         **end for**
7:         {Phase 2: Clustering}
8:         $shingle\_list \leftarrow$ Sort $shingle\_list$ by shingle hash
9:         Create a list of <content-ID, content-ID, count = 1> by expanding $shingle\_list$, which is denoted by L.
10:        $L \leftarrow$ Sort L by first content-ID and second content-ID
11:        Merge L (adding the counts for matching <content-ID, content-ID, count>).
12:        Initiate undirected graph G with all the segments as its vertices and with no edges.
13:        **for** each pair p in L **do**
14:            Calculate jaccard similarity coefficient j of p.
15:            **if** j exceeds the threshold F1 **then**
16:                Add a edge (weight=j) between the two segments.
17:            **end if**
18:        **end for**
19:        Find all the connected components C_G by BFS.
20:        {Phase 3: Detecting template clusters}
21:        **for** j=1 to the number of connected components in C_G **do**
22:            $g \leftarrow$ the gravity-segment of $C_G[j]$, where the sum of weights of edges linked with gravity-segment is maximum.
23:            **if** ClusterPro$(C_G[j])$>F2 **then**
24:                Set $C_G[j]$ is $template$;
25:            **else**
26:                Set $C_G[j]$ is $informative$;
27:            **end if**
28:        **end for**
29:        {Phase 4: Constructing classifier}
30:        Create a 0-1 Classifier based on all the gravity-segments in S.
31:    **end for**
32: **end procedure**

4.2   Constructing Template

Through SSOM tree, segments with the common presentation style can be aggregated together. Each node S of SSOM is a document collection, which contains many segments from different pages.

The second step of our algorithm is constructing a template from the SSOM tree. $ConstructTemplate$ algorithm is described in Algorithm 4, which is divided into four phases as follows:

First, we produce a list of <shingle-hash, content-ID> pairs by shingling algorithm, where a shingle hash is the hash value of a shingle and a content-ID is the unique identification of the content the shingle appears in.

Second, for each node, we cluster its segments by computing the jaccard similarity coefficient of each two segments.

Third, we identify *template* clusters by checking if the possibility is larger than the threshold F2.

At last, for each node, we create a 0-1 classifier which can be used to classify a segment by comparing the similarity of the gravity of each cluster and a given segment.

Shingling. In phase 1 of the *ConstructTemplate* algorithm, we use the shingling algorithm. In our method, the process of clustering the segments in phase 2 is based on the similarity. Here we use shingles of segment to represent those segments. As mentioned above, for each SSOM node S, $S.content$ is a set of text-contents of segments which comes from different pages. A procedure of shingling is performed on the text-contents: a window of size W is slid over each text-content in $S.content$ and a list of <shingle-hash, content-ID> pairs are obtained by combining the hash value of each shingle and the ID of content it appears in. In our experiments, W is set to 2 empirically.

Clustering. The second phase of *ConstructTemplate* algorithm is clustering the segments. In this paper, we use a shingling-based clustering technique to determine almost-similarities, which is based on the method Broder et al. proposed in [2], for this method has been proved to be highly efficient. The detail of our method is described as follows:

First, we sort the shingle list we have obtained in phase 1 by shingle-hash values. The result is presented by the list L of <shingle-hash, content-ID> pairs (Line 3-5).

After that, we generate a list of all the pairs of contents that share any shingles, along with the number of shingles they have in common. To do this, we expand L into a list of <content-ID, content-ID, count of common shingles> triplets by taking each shingle that appears in multiple contents and generating the complete set of <content-ID, content-ID, 1> triplets for that shingle. We then apply the sort and merge procedure (addling the counts from matching content-ID C content-ID pairs) to produce a list of all <content-ID, content-ID, count> triplets sorted by the first content-ID and the second content-ID. The process is described in Lines 7-10.

Then, we create an undirected graph G with all the contents as its vertices and no edges. Every shingle in the content represents as its feature, and two segments are linked and the weight is set as j by checking if the jaccard similarity coefficient j is larger than the threshold F1, which is set to 0.4 by our experiment. (Lines 11-17). The formula of calculating jaccard similarity coefficient is shown as follow.

$$j(x,y) = \frac{count}{num(content_x) + num(content_y) - count} \tag{1}$$

where *count* is the number of common shingles of *segment_x* and *segment_y*, and *num(content_i)* is the number of shingles of *segment_i*. At last, we find all connected components C_G by BFS (Breadth First Search) algorithm.

Identifying Template Clusters. Intuitively, segments with the similar contents and common look and feel from different pages tend to templates. The more similar segments are, the more likely these segments are templates. In our work, we consider the node S is template based on the following assumptions: (1) the less groups that S has and the less balanced that the groups are, the more possible S is template; (2) the larger the probability of a cluster C in S is, the more possible C is template. Information theory (or entropy) and probability are intuitive choice.

Definition 3 (Node probability): For an element node S in the SSOM, let $C_G[k]$ is the *kth* cluster of C_G, m be the number of segments in S and P_K is the probability that $C_G[k]$ appears in C_G. The node probability that S is template is defined by

$$NodePro(S) = \begin{cases} 1 - |\sum_{k=1}^{|C_G|} P_k * log_m P_k| & if \quad m > 1 \\ 0 & if \quad m = 1 \end{cases} \qquad (2)$$

Definition 4 (Cluster probability): For a cluster C in C_G, we consider it template through its probability and node probability. The cluster probability that C is template is defined by

$$ClusterPro(C_G[i]) = r * P_i + (1 - r) * NodePro(S) \qquad (3)$$

In (3) P_i denotes the probability that $C_G[i]$ appears in C_G, and r is the attenuating factor which is set to 0.5.

For each cluster C in node S, we set C is *template* if $ClusterPro(C)$ exceeds the threshold F2, which is set to 0.5.

Creating 0-1 Classifier. The last phase of *ConstructTemplate* algorithm is creating template-classifier. For classification, we use the k-Nearest Neighbor classifier (KNN), which performs well at here. For each cluster C, $C.gravitie$ is set as X of a training data and its result gained in phase 3 as Y. Given a new segment s, we can find the most *nth* nearest clusters and classify s into *template* by checking if the number of *template* clusters is larger than the number of *informative* clusters.

Identifying *template* clusters and creating template-classifiers for all the nodes in the SSOM tree can be easily done by traversing the SSOM tree, and therefore we will not discuss it further more.

4.3   Detecting Template

Here we present the last step of template detection algorithm, as shown in Algorithm 5. Given a new page β, firstly we parse it into SOM tree. And then we

---

Algorithm 5. DetectTemplate

Input:
  • The root SSOM node: S
  • The root SOM node: N
Output:
  • All template segments: T
1: **procedure** :
2:     **if** $S.label == N.label$ **then**
3:         **if** S is a leaf node **then**
4:             Add the descendant nodes into N.
5:         **end if**
6:         Consider N as a segment seg.
7:         Classify the seg by $S.classifier$.
8:         **if** $S.classifier$ classifies seg into $template$ **then**
9:             $T \leftarrow seg$
10:         **end if**
11:         **for** each child n in N **do**
12:             **if** $\exists s \epsilon S.children$ and $n.label=s.label$ **then**
13:                 $T \leftarrow T \cup$ DetectTemplate(s, n);
14:             **end if**
15:         **end for**
16:     **end if**
17:     Return T;
18: **end procedure**

---

can detect all the template segments by mapping and classifying every segments recursively.

5   Experiments

This section evaluates our proposed informative contents discovering algorithm. Since the main purpose of Web content extraction is to improve Web data mining, we performed two data mining tasks, i.e., clustering and classification, to test our system. To validate our method, we compared our approach with these methods, i.e., Origin (results before cleaning), DOM Based Algorithm [7], Text Based Algorithm [7], SST Based Algorithm [16] [15] and RTDM Based Algorithm [13]. The parameters of these methods are set to the best value in line with the corresponding experiments:

  • DOM Based Algorithm with parameters (F=100);
  • Text Based Algorithm with parameters (W=32, F=50, D=0, P=1000);
  • RTDM Based Algorithm;
  • SST Based Algorithm with parameters (r=0.9, t=0.4);

In the following subsections, we first describe the data sets and evaluation measures used in our experiments. After that, we present the results of the experiments, and also give some discussions.

Table 1. Number of Web pages and their classes

Web sites	PCConnection	Amazon	CNet	J&R	PCMag	ZDnet
Notebook	560	410	431	60	145	198
Camera	156	230	206	150	138	139
Mobile	20	36	42	32	47	108
Printer	423	610	123	127	110	89
TV	267	589	146	171	56	72

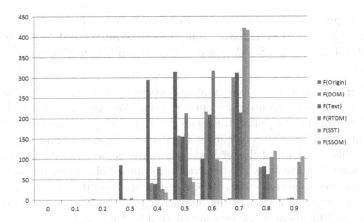

Fig. 3. Distribution of F score

5.1 Data Sets and Evaluation Measures

In this paper, we crawled six distinct commercial Web sites: PCConnection[1], Amazon[2], CNet[3], J&R[4], PCMag[5] and ZDnet[6]. The six Web sites contain Web pages of many categories or classes of products. We choose the Web pages that focus on the following categories of products: Notebook, Camera, Mobile, Printer and TV. Table 1 lists the number of documents downloaded from each Web site, and their corresponding classes.

Since we test our method using clustering and classification, we use the popular F score measure to evaluate the results for clustering and classification of our method and the baselines. F score is defined as follows:

$$F = \frac{2P*R}{P+R}$$

where P is the precision and R is the recall. *Fscore* is the weighted harmonic mean of P and R, which reflects the average effect of both precision and recall.

---

[1] http://www.pcconnection.com/

[2] http://www.amazon.com/

[3] http://www.cnet.com/

[4] http://www.jandr.com/

[5] http://www.pcmag.com/

[6] http://www.zdnet.com/

Table 2. The average F scores

Method	F(Origin)	F(DOM)	F(Text)	F(RTDM)	F(SST)	F(SSOM)
Avg	0.455	0.6275	0.631	0.583	0.701	**0.712**

5.2   Experimental Results

We now present the experimental results of Web page clustering and classification.

In this paper, we constuct the SSOM of each Web site by all the crawled pages for each site. After labelling sentences of sampled 100 Web pages, the entropy-threshold of each SSOM node is determined automatically.

Clustering. We stress the experimental procedure we use is the same as in [9], but the sets of pages used are not the same. We use the popular k-means clustering algorithm. We put all the 5 categories of Web pages into a big set, and use the clustering algorithm to cluster them into 5 clusters. Since the k-means algorithm selects the initial cluster seeds randomly, we performed a large number of experiments (800) to show the behaviors of k-means clustering on our method and the baselines. Figure 3 shows 10 bins of F score from 0 to 1 with an interval as 0.1 and gives the statistics of the number of experiments whose F scores fall into each bin. The average F scores are plotted in Table 2.

F(Origin) represents the F score of clustering based on original noisy Web pages; F(DOM) represents the F score of clustering based on DOM-Based Algorithm [7]; F(Text) represents the F score of clustering based on Text-Based Algorithm [7]; F(RTDM) represents the F score of clustering based on RTDM Based Algorithm [13]; F(SST) represents the F score of clustering based on SST-Based Algorithm [9]; F(SSOM) represents the F score of clustering based on our method.

Thus, we can clearly observe that clustering results based on all the cleaning methods are dramatically better than the results using the original noisy Web pages. Our method also produce a better result than the other baselines for Web page clustering.

Classification. The evaluation of impact our method over classification also follows the experimental procedure proposed in [9]. For classification, we use the Naive Bayesian classifer (NB), which has been shown to perform very well in practice by many researches. We experiment with three different configurations for classification tasks using all possible pairs of product categories. The configurations are summarized in Table 3. For each pair, we train the NB classifier using a Training Set and then run the classifier over the corresponding Test Set.

In Table 4, we can observe that classification results based on all the cleaning methods are dramatically better than the results using the original noisy Web pages. Our method produce a little better result than SST and much better than the other three baselines for Web page classification.

Table 3. Configuration of classification experiments

Configuration	Training Set	Test Set
1	Pages from categories p and q from site i	Pages from categories p and q from all sites except i
2	Pages from category p from site i and pages from category q from site $j \neq i$	Pages from categories p and q from all sites except i and j
3	Pages from category p from site i and pages from category q from site $j \neq i$	Pages from categories p and q from all sites except pages in the Training Set

Table 4. F-measure of classification results for each configuration

Categories		Conf.1						Conf.2						Conf.2					
p	q	Orig	DOM	Text	RT	SST	SSOM	Orig	DOM	Text	RT	SST	SSOM	Orig	DOM	Text	RT	SST	SSOM
Not	Cam	0.901	0.965	0.971	0.976	**0.981**	0.976	0.620	0.832	0.847	0.853	**0.877**	0.873	0.469	0.655	0.678	0.685	**0.764**	0.758
Not	Mob	0.852	0.921	0.905	0.934	0.917	**0.935**	0.504	0.773	0.789	0.784	0.799	**0.824**	0.410	0.509	0.534	0.513	0.558	**0.560**
Not	Pri	0.901	0.987	0.989	0.988	0.979	**0.991**	0.618	0.852	0.864	0.848	0.856	**0.867**	0.467	0.651	0.702	0.664	0.783	**0.793**
Not	TV	0.900	0.958	0.961	0.985	**0.996**	0.987	0.592	0.802	0.813	0.816	0.862	**0.868**	0.441	0.679	0.753	0.651	**0.795**	0.789
Cam	Mob	0.854	0.899	0.902	0.901	**0.913**	0.907	0.531	0.868	0.897	0.882	**0.959**	0.929	0.428	0.657	0.732	0.637	**0.776**	0.764
Cam	Pri	0.876	0.904	0.943	0.963	0.971	**0.978**	0.667	0.821	0.834	0.865	0.873	**0.881**	0.504	0.701	0.746	0.699	0.805	**0.809**
Cam	TV	0.813	0.898	0.912	0.974	0.982	**0.990**	0.621	0.836	0.857	0.869	0.904	**0.925**	0.478	0.689	0.733	0.682	0.801	**0.814**
Mob	Pri	0.874	0.901	0.907	0.902	**0.914**	0.912	0.634	0.819	0.834	0.822	**0.880**	0.872	0.489	0.663	0.764	0.634	**0.792**	0.782
Mob	TV	0.783	0.896	0.947	0.870	0.968	**0.973**	0.569	0.855	0.928	0.802	0.947	**0.948**	0.442	0.631	0.768	0.624	0.784	**0.790**
Pri	TV	0.897	0.978	0.983	0.988	**0.996**	0.992	0.583	0.863	0.871	0.855	**0.906**	0.905	0.432	0.775	0.792	0.721	0.819	**0.826**
Avg		0.865	0.931	0.942	0.948	0.962	**0.964**	0.594	0.832	0.853	0.840	0.886	**0.889**	0.456	0.661	0.720	0.651	0.768	**0.768**

6 Conclusions

In this paper, we propose a novel algorithm for template detection which is based on SSOM (Site-oriented Segment Object Model). Given a Web site, a SSOM tree is constructed from a collection of pages in the site. Then a method based on clustering and information theory is used to detect templates and create template-classifiers. After that, the SSOM tree can be used to detect templates of Web pages in the Web site with mapping and classifying operation.

Experiments conducted on 6 Web sites show that our approach produces quite strong result compared with other five typical algorithms (including original texts).

In future work, we will investigate the application of our method to the processing of short-text data, such as forum and micro blog dataset. Further, we plan to explore the improvement on how to discovery informative contents of a large scale of Web pages automatically.

References

1. Bar-Yossef, Z., Rajagopalan, S.: Template detection via data mining and its applications. In: Proceedings of the 11th International Conference on World Wide Web, pp. 580–591. ACM (2002)
2. Broder, A.Z., Glassman, S.C., Manasse, M.S., Zweig, G.: Syntactic clustering of the web. Computer Networks and ISDN Systems 29(8), 1157–1166 (1997)

3. Cai, D., Yu, S., Wen, J.R., Ma, W.Y.: Vips: a vision-based page segmentation algorithm. Tech. rep., Microsoft technical report, MSR-TR-2003-79 (2003)
4. Chakrabarti, D., Kumar, R., Punera, K.: Page-level template detection via isotonic smoothing. In: Proceedings of the 16th International Conference on World Wide Web, pp. 61–70. ACM (2007)
5. Debnath, S., Mitra, P., Pal, N., Giles, C.L.: Automatic identification of informative sections of web pages. IEEE Transactions on Knowledge and Data Engineering 17(9), 1233–1246 (2005)
6. Fernandes, D., de Moura, E.S., da Silva, A.S., Ribeiro-Neto, B., Braga, E.: A site oriented method for segmenting web pages. In: Proceedings of the 34th International ACM SIGIR Conference on Research and Development in Information Retrieval, pp. 215–224. ACM (2011)
7. Gibson, D., Punera, K., Tomkins, A.: The volume and evolution of web page templates. In: Special Interest Tracks and Posters of the 14th International Conference on World Wide Web, pp. 830–839. ACM (2005)
8. Kao, H.Y., Chen, M.S., Lin, S.H., Ho, J.M.: Entropy-based link analysis for mining web informative structures. In: Proceedings of the Eleventh International Conference on Information and Knowledge Management, pp. 574–581. ACM (2002)
9. Lin, S.H., Ho, J.M.: Discovering informative content blocks from web documents. In: Proceedings of the Eighth ACM SIGKDD International Conference on Knowledge Discovery and Data Mining, pp. 588–593. ACM (2002)
10. Ma, L., Goharian, N., Chowdhury, A., Chung, M.: Extracting unstructured data from template generated web documents. In: Proceedings of the Twelfth International Conference on Information and Knowledge Management, pp. 512–515. ACM (2003)
11. Reis, D.D.C., Golgher, P.B., Silva, A., Laender, A.: Automatic web news extraction using tree edit distance. In: Proceedings of the 13th International Conference on World Wide Web, pp. 502–511. ACM (2004)
12. Song, R., Liu, H., Wen, J.R., Ma, W.Y.: Learning block importance models for web pages. In: Proceedings of the 13th International Conference on World Wide Web, pp. 203–211. ACM (2004)
13. Vieira, K., da Silva, A.S., Pinto, N., de Moura, E.S., Cavalcanti, J., Freire, J.: A fast and robust method for web page template detection and removal. In: Proceedings of the 15th ACM International Conference on Information and Knowledge Management, pp. 258–267. ACM (2006)
14. Wang, J., Lochovsky, F.H.: Data-rich section extraction from html pages. In: Proceedings of the Third International Conference on Web Information Systems Engineering, WISE 2002, pp. 313–322. IEEE (2002)
15. Yi, L., Liu, B.: Web page cleaning for web mining through feature weighting. In: International Joint Conference on Artificial Intelligence, vol. 18, pp. 43–50. Lawrence Erlbaum Associates Ltd. (2003)
16. Yi, L., Liu, B., Li, X.: Eliminating noisy information in web pages for data mining. In: Proceedings of the Ninth ACM SIGKDD International Conference on Knowledge Discovery and Data Mining, pp. 296–305. ACM (2003)

Analysis and Evaluation of Web Pages Classification Techniques for Inappropriate Content Blocking

Igor Kotenko[1,2], Andrey Chechulin[1], Andrey Shorov[1,3], and Dmitry Komashinsky[4]

[1] St. Petersburg Institute for Informatics and Automation,
39, 14th Liniya, Saint-Petersburg, Russia
{ivkote,chechulin,shorov}@comsec.spb.ru
[2] St. Petersburg National Research University of Information Technologies,
Mechanics and Optics, 49, Kronverkskiy Prospekt, Saint-Petersburg, Russia
[3] Saint-Petersburg Electrotechnical University "LETI",
Professora Popova str. 5, Saint-Petersburg, Russia
[4] F-Secure Corporation, Tammasaarenkatu 7, PL 24, 00181 Helsinki, Finland
dmitriy.komashinskiy@f-secure.com

Abstract. The paper considers the problem of automated categorization of web sites for systems used to block web pages that contain inappropriate content. In the paper we applied the techniques of analysis of the text, html tags, URL addresses and other information using Machine Learning and Data Mining methods. Besides that, techniques of analysis of sites that provide information in different languages are suggested. Architecture and algorithms of the system for collecting, storing and analyzing data required for classification of sites are presented. Results of experiments on analysis of web sites' correspondence to different categories are given. Evaluation of the classification quality is performed. The classification system developed as a result of this work is implemented in F-Secure mass production systems performing analysis of web content.

Keywords: Classification of web pages, data mining, text analysis, HTML structure analysis, analysis of URL addresses, hierarchical classification.

1 Introduction

Nowadays, the Internet is one of the main ways of information obtaining. Lack of mechanisms that may classify information and control access to it in the Internet creates a problem of accessing unacceptable information by a certain circle of people. First of all, it concerns the need to restrict access to certain types of information according to age categories. In addition, limiting of automatic access to sites belonging to the higher risk categories (for example, "adult sites", "sites with unlicensed software", etc.) will increase security of users from malicious and unwanted software.

Contemporary web resources have a complex hierarchical structure and consist of multiple elements, including formatted text and graphical content, program code and links. This causes a number of problems inherent to the task of classifying web pages, with necessity to analyze colossal volumes of heterogeneous, often conflicting and

P. Perner (Ed.): ICDM 2014, LNAI 8557, pp. 39–54, 2014.

changing data. The main purpose of the presented work was to develop a classification system intended to block inappropriate content for deploying in F-Secure [8] software systems performing analysis of web content. The main theoretical contribution of the paper is to develop models, techniques and algorithms for web page classification based on data mining and machine learning. The developed models and techniques include generalizations of binary classifiers of web pages based on text information, the text content of individual structural elements (HTML-tags), their address information (Universal Resource Locator, URL), and combining of binary classifiers for obtaining an unequivocal classification result. Within the framework of the developed models and algorithms, there was created a system for classification of web pages based on text data, whose language is different from the language in which binary classifiers were trained (the training language is English, the languages of recognition are the main European languages, Chinese and others). Developed models, techniques and algorithms made it possible to build an automated system for classification of web sites, allowing to collect and process data required for training and testing of classifiers. A tool that uses trained classifier for rapid classification of web pages in different languages was developed.

The novelty of the research presented in the paper is in the integrated use of existing and modified approaches to classification of text and other information to determine the web page category. In addition, a new approach to the construction of dictionaries of text attributes was used at construction of new features that allowed enhancing the overall quality of the classification. The paper presents main elements of the research and development performed; its structure is organized as follows. *Second section* discusses the main results of relevant research. *Third section* provides a general description of the developed approach and the basic tasks solved during its development. Special attention is paid to the system characteristics used in the learning process. Essential aspects of procedures of receiving and processing data during the system training phase are presented in *fourth section*. Brief description of software implementation of the approach and the main results, causing selected decisions on the structure of decision-making procedures and organization of the system are described in *fifth and sixth sections* respectively. The main conclusions and further research directions are discussed in *seventh section*.

2   Related Work

The basic list of problems inherent to this subject area and approaches to their solution is presented in the review by Qi and Davison [24]. It summarizes the state of research at the end of the first decade of the XXI century. It discusses the possible variants for setting web page classification problem (binary, multiclass, hierarchical) and the main models for representation of entities used in the Internet as a set of interrelated elements. Differences of classification tasks for web pages and text are determined. On this basis, the list of features that are applicable to classification is specified. The basic applicable models for web page representation and special methods that use them are considered.

Let us consider the papers published after the appearance of this review. Shibu and others [27] considered the optimization issues of learning process in this class of systems due to the combined use of traditional procedures for selection of significant features with data Page Rank. Patil [22] investigated the applicability of Naive Bayes (NB) classifier for learning of web page classification systems within the individual groups of internal features of HTML documents. For classification of web pages Xu et al. [29] proposed the algorithm called Link Information Categorization (LIC), based on the k nearest neighbors (kNN) method. Its essence lies in the definition of the category of the classified web page based on analyzing links that other web pages make to this one. Calculation of relation level of web pages containing links to a particular category is done. The classification is performed by nine categories.

In general, it should be noted that most often used features that are applied for web page classification are extracted from the page text content. For instance, Dumais and Chen [6] separated concepts of web page text, header information and descriptive information service tag "meta". They implemented the Support Vector Machine (SVM) method. Lai and Wu [18] used two approaches to obtain necessary features for classification: meaningful term extraction and discriminative term selection. The main idea is to extract unknown words and phrases that belong to specific domain. Thus obtained terms are very specific to a particular domain. The classification was made according to five categories. Vector space model [26] was used for the classification.

Tsukada et al. [28] performed web pages classification by analyzing the nouns extracted from web pages. For removal of common words that are not related to a particular topic, and auxiliary words a stop-list was used. Choice of words needed for classifier training was implemented using the Apriori algorithm [1]. The method of Decision Trees (DT) was used for classification. Training was carried out on a set of objects belonging to five categories. In the work of Kwon et al. [16, 17] text features, extracted from different blocks of formatted text web pages, had three different levels of significance. Solution of the classification problem was achieved using the kNN.

Qi et al. [23] chose as the source of textual data not only the web page, but also its "neighbors", represented by the terms "parent", "descendant", "brother", etc. SVM and NB methods were used. An important result of this work is the demonstrated improvement of classification using data from neighboring web pages, which, however, leads to additional computational costs.

An alternative direction of research in the field of classification of web pages is the usage of text data extracted from their address (URL). Baykan et al. [2] performed search of tokens (i.e. search of individual words in the URL using the dictionary) and their decomposition into sequences of length from 4 to 8 characters to obtain mined features. Furthermore, the combination of all received n-grams, regardless of their length, was used. Classifiers models were based on SVM, NB and the maximum entropy method (MEM). Classification of web pages by URL was also realized in several papers of Kan et al. [11, 12]. The following features were implemented for training of classifiers: tokens, URL elements, structural patterns, association of abbreviations with full words combinations. SVM and MEM were used for classifier training.

Among relevant publications that explore aspects of SVM for web page classification we also mark works of Joachims [10], Dumais et al. [7] and Yang et al. [31]. Yang et al. also apply kNN, used by Calado et al. [3] and Lam et al. [19]. More information about applying NB can be found in the works by already mentioned Joachims [10], Dumais et al. [7], as well as Lewis [20], Chakrabarti et al. [4] and McCallum et al. [21].

3   General Approach to Websites Classification

3.1   Classification and Features Selection

For classifying the content of web pages we must take into consideration a number of features and particularities peculiar to the presentation of content in the Internet. Sources of information are web page addresses (URL) themselves, text content, text structure (HTML, Hyper Text Markup Language), structure of links, showing "location" in relation to other elements of web, and internal multimedia content [24].

For most ways of web page representation the source text information that can be extracted from its address (URL) and individual formatted elements of its content (HTML tags) as well as graphic information (such as images) are the most informative sources. In addition, historical information about the web page changes, when it is available, is of great importance.

Typically, in the first place the web content classification systems use the text displayed by web pages. The obtained results are applied to clarify the next steps of the decision process. The proposed approach to training is based on the application of machine learning for obtaining classifiers necessary to build a common classification system. Using the information obtained during raw data processing, the lists of major keywords for certain categories (based on the structural features) are formed. Direct classifiers training occurs on the base of text data obtained in the process of analysis of web pages belonging to predetermined topics. Preliminary data analysis showed that in some cases the conflict situations requiring resolution at the stage of preparing the input data for training are possible. An example of this conflict is the proximity of the text content of certain web pages, directly related to pharmacology, to the domain of drugs. Obviously, this may lead to errors such as "false positive" that stipulates wrong web page domain identification. The proposed solution to this problem is to break the thematic data sets into a number of categories, specifying one or another aspect of the topic. For example, a topic related to alcoholic beverages can be broken down into categories that reveal issues brewing, wine making and so on.

In the paper we consider the web pages belonging to a number of categories related to the following topics: "adult", "alcohol", "gambling", "tobacco", "dating", "drugs", "hate", "violence", "weapon", "religion", "occults". For descriptions of other topics and related categories we used generalized notion of "unknown".

3.2   Construction of Base and Combining Classifiers

We rely on the formal description of the problem of knowledge extraction from the Web [5] refined by Kleinberg et al. [13]. Based on the data available from the web page address (URL) and the formatted text content (HTML), a three level decision scheme was developed.

Formal Statement of the Problem. It is given a set O of objects $\{o_1, o_2, ..., o_n\}$ and the set C of labels of target classes $\{c_1, c_2, ..., c_k\}$. Each object $o_i \in O$ can be represented by the corresponding element x_i of the set X of object descriptions, where x_i is a set $[x_{i,1}, x_{i,2}, ..., x_{i,m}]$. Getting object descriptions is achieved by the conversion $f_{Map}: O \rightarrow X$, defined by applying a set of features $F = \{f_1, f_2, ..., f_m\}$, each element of which is a transformation of the form $f: O \rightarrow V$ which puts for an element of the set of objects O an element of characteristic values V. There exists a dependence $f_{Target}: X \rightarrow C$, assigning to each object description x_i of the object o_i an element of class labels $c_j \in C$. It is necessary to construct an algorithm f_{Result}, approximating the target dependence f_{Target} on an accessible collection of objects $O_{Training} \in O$ and on the whole set of objects O.

Description of the Approach. Let h be the number of individual aspects, within the frames of which we study each element of the set O, and z is the number of features used to describe the aspects inherent in each of the designated categories, under a separate aspect. Then the total number m of elements of set F is equal to $k \times h \times z$, and the set of features is represented as $F = U_{l \in (1,h), j \in (1,z)} F_{l,j}$.

Prediction of the first level $p1_{i,l,j}$ about the belonging of description of the first level x_i of the object o_i to the category j under the aspect l is the result of applying the algorithm of the first level $a1_{l,j}: x_i \rightarrow P1_{i,l,j}$, $p1_{i,l,j} \in P1_{i,l,j}$, $P1_{i,l,j} = \{true, false\}$.

Description $d2_{i,l}$ of the second level of the object o_i within the aspect l is represented as a set $[p1_{i,l,1}, p1_{i,l,2}, ..., p1_{i,l,k}]$ of predictions of the first level and characterizes the object belonging to each of the categories of the set C as a separate aspect.

Prediction $p2_{i,l}$ of the second level about belonging the description $d2_{i,l}$ to one of the target categories within the aspect l is the result of applying the algorithm of the second level $a2_l: d2_{i,l} \rightarrow C$.

Description $d3_i$ of the third level of the object o_i is represented as a set $[p2_{i,1}, p2_{i,2}, ..., p2_{i,h}]$ of predictions of the second level and characterizes the object belonging to a category of the set C throughout the whole set A of aspects.

Prediction $p3_i$ of the third level on belonging the description $d3_i$ to one of the categories is the result of applying the algorithm of the third level $a3: d3_i \rightarrow C$.

Within the framework of the presented approach the development of an algorithm f_{Result} requires formation of a set of algorithms of the first, second and third levels.

Used Metrics. The degree of proximity f_{Result} and f_{Target} is assessed based on the F-measure metric that combines measures of accuracy and completeness of the solutions, which are in turn the characteristics determined by the errors' index of the first and second kind (false-positives and false-negative checks, respectively): $F_\beta = ((1 + \beta^2) \cdot tp)/((1 + \beta^2) \cdot tp + \beta^2 \cdot fn + fp)$, where tp is the number of correctly identified cases for a target category, fn is the number of false-negatives, fp is the number of false-positives, β is a coefficient of significance of false-negatives comparing with false positives (within the frames of the results presented here, both types of errors are equal, in order to keep focusing on FP-proof solutions $\beta \ll 1$).

The developed decision making scheme is represented in Fig. 1. Elements of the first level are functional blocks oriented to certain categories. They use pre-trained classifiers giving judgments about whether a particular vector, characterizing the specific web page, refers to a certain category (thus the elements of the first level of the developed scheme solve the problem of binary classification). Each classifier is formed based on descriptions of objects in space of features inherent to the target domain (category). The need for this is confirmed by obvious use of specific terms and phrases within individual topics and categories. Thus, the key feature of the scheme at this level is the use of binary classifiers focused on individual categories and features inherent to them. The resulting solutions are hereinafter called as the first level predictions. Elements of the second level of the developed scheme use classifiers oriented for making decision about belonging the vector of descriptions of the given web page to one of the given domains (categories). Here the information obtained in the analysis of individual structural aspects of a web page (for example, data from web page address or its elements' formatting) is used. Descriptions going to the input of the aspect-oriented elements are formed by prediction values of the first level. The resulting solutions are hereinafter called as the second level predictions.

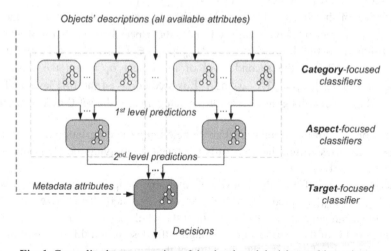

Fig. 1. Generalized representation of the developed decision making scheme

Predictions of the second level are used to generate descriptions of the analyzed web pages that are used for training and decision making by element of the third level, intended to form the final decisions.

Using this classification scheme was due to the decision of a number of research tasks, outlined in details in [14-15]. Firstly, we received a solution on the choice of the feature space used for training classifiers that are used by first level elements. It was shown that the use of relatively small sets of features relating to the target category is sufficient to create functional blocks of the scheme. Secondly, a set of experiments aimed at evaluation of the applicability of different groups of features used to train the classifiers of the second and third levels of the scheme was performed. The result of this work was the decision to use at these levels only the solutions of the first level elements without involving the initial set of features.

To clarify the decision-making process, at the third level of the scheme in addition to the second level predictions, the information describing the context of obtaining web page data and general characteristics of its volume, origin, etc. may be used.

The proposed approach allows training a classifier consisting of atomic classifiers of different levels, specializing in working with certain types of data. Separate particularity of the approach is the ability to make quick update and supplement of the decision making scheme at entering new categories and domains or the emergence of new data sets useful for retraining already used decision making elements.

4   Obtaining and Processing of Data

Obtaining of raw data and their preparation for further analysis is one of the most important and effort consuming steps. This stage includes primary data loading, selection of necessary data, their retrieval from the web page and formation of combined features based on existing ones. This stage can be divided into the following sub-steps: (1) creating a list of web pages to build the training and test samples; (2) download and parsing of the contents of web pages; (3) extraction and formation of combined features by web pages content. As sources for the classification the following aspects of the data specifying a web page were selected: URL identifier, full text contained in the page and visible to the user, some text areas, marked with most thematically correlative tags and general statistics of tags.

In the research process the SVM, DT and NB classifiers were used. For each of the classifiers different combinations of classification parameter values were considered.

Let us consider different source data types in more details.

4.1   Processing and Analysis of Textual Information

The main source of data for the construction and performance of the trained model is the general web page text and the text contained in specific tags. To perform the analysis of the text by using the classifier, the text is presented in the form of keywords or phrases (wordforms) based on different methods of decomposition of the text.

There are many ways to split the text into individual wordforms. During performing this research the analysis of the following methods was done: (1) separating at appearance of non-alphanumeric characters, (2) separation into clauses on the basis of punctuation analysis, and (3) division into tokens (based on the dictionary of Rapid Miner [25]).

We also analyzed the following text preprocessing techniques: (1) stemming (process of finding a basis for a given word of the original word); (2) replacement of individual words with their hyponyms (based on the Oxford Dictionary); (3) replacement of individual words with their hyperonyms (based on the Oxford Dictionary).

In addition, we built a list of prepositions, conjunctions, pronouns, etc. that do not have any independent meaning, and in the case of the analysis of individual words the elements of this list were excluded from processing.

Let us consider in more detail the technique of building the vocabulary of basic wordforms for further training of the classifier. This technique is based on the analysis of the frequency characteristics. It involves the following steps: (1) Preparation of text content for all web pages (here we can use both the general text and the text stored in individual tags). (2) Exception of stop words from all texts (i.e. words that do not carry the semantic load). Efficiency of the inclusion of this step was checked in experiments. (3) Processing of all texts by stemmer (i.e. separation of the part that is the same for all the grammatical forms of a word by cutting off endings and suffixes). Efficiency of the inclusion of this step was checked in experiments. (4) Creation of the lists of keywords for each category with given values of TF (term frequency), where TF is the value calculated as the ratio of the number of occurrences of a word to the total amount of words in all web pages of the category, defining the importance of the words within a category. (5) Calculation of IDF (inverse document frequency) that is inversion of frequency with which a word occurs in the selected categories. Using the IDF allows to create the word vector of high importance within one category and rarely found in other categories.

When forming the dictionary we also used an original technique for calculation of TF and IDF. TF is normally calculated as the ratio of the number of occurrences of a word to the total number of words in the document. In our research, a category was used in the role of a document, and the number of occurrences was determined as the number of web pages containing this word. At the same time we investigated various indicators characterizing the presence of words in the text (for example, a single occurrence of the word in the text, consisting of 100 thousand words, does not mean that the text is relevant to the category to which this word belongs). The same approach was used for IDF, i.e. the word having low value of TF for some category may be considered as absent in this category. In experiments we checked different boundary parameters determining occurrence of a word in a text or a category, and found their optimal values.

After completion of the previous stage, for each web page a table is filled in which a value from the set $\{0,1\}$ corresponds to each wordform. This value determines whether this web page specific contains a wordform belonging to a category.

In the next step the training of atomic classifiers for particular categories is performed. Each classifier receives on the input a limited number of wordforms that are the most relevant to one category or another. Thus, for training the classifier, the following source data for each web page were used: (1) the number of occurrences of each wordform from the dictionary in the text of the web page (in the experiments we considered different sizes of dictionaries), (2) the category of the web page.

After comparing the quality of text classification using various techniques we chose the technique based on decision trees, as it showed high level of precision along with adequate recall, accuracy and F-measure.

4.2   Processing and Analysis of URL Addresses of Web Pages

To analyze the URL addresses of web pages we used a technique similar to the approach used in the analysis of textual information. However, given the particularities of URL, the selection of individual elements of the text was performed with a method of finding n-grams. After that for these n-grams we used the same steps as for the wordforms in text analysis (TF-IDF).

Let us consider in more details the technique for building the dictionary of n-grams for further training of the classifier. This technique comprises the following steps: (1) collection of URL addresses of all web pages belonging to the same category in one text object (thus, for each category a separate text object was formed); (2) formation of text tokens based on special characters that are contained in URL addresses; (3) removing of stop words and applying algorithms of stemming to obtained text tokens; (4) generation of n-grams for processed lists of tokens. Thus, for training a classifier the following input data for each web page were used: presence of each n-gram from dictionary in URL address of the web page; integer attribute that characterizes the total number of n-grams presented in the URL address of the web page and specific to a particular category; web page category.

4.3   Processing and Analysis of Information on Structure of Web Pages

To identify such categories as news, forums, blogs, etc. an approach was used, based on the analysis of the HTML structure of web pages. An important peculiarity of these categories is heterogeneous text content and attributes contained in the structure of tags (number of references, tables, headers, etc.). To select tags that allow to identify web pages of certain categories, the information weight of each tag was calculated, and the sample of tags, based on a predetermined boundary, was produced. For training of the classifier the following input data for each web page were used: (1) the number of occurrences of each of the selected tags in the HTML code of the web page, (2) the length of the text in different tags, and (3) the category of the web page.

4.4   Classification of Sites in Languages Other Than English

To classify sites in foreign languages it is suggested to use translation of text content of web pages from the original language into the language that was used for training of classifiers. This approach allows to classify web pages in any language supported by the system of automatic translation, using models of classifiers trained on English web pages. There is no need to prepare additional classifiers to categorize web pages for each specific language. This eliminates the need for preparation of additional training data sets and further maintenance of them in up to date state. In its turn, the use of classifiers focused on a specific language allows to use different grammatical structures for training that can improve the classification quality.

4.5   Other Sources of Information about Web Pages

In the research we also considered other sources of data based on which it is possible to get information characterizing the category to which the web page belongs.

For example, the following sources were analyzed: (1) data about web pages that are referenced by the analyzed web page, or which in turn refer to it; (2) information from WhoIs servers; (3) existing lists of web pages containing those labels (for example, lists used in parental control systems); (4) defined category of the web page for a certain period of time (it allows to take into account changes in the web site content).

The mentioned sources currently do not participate in the process of classification for the following reasons: (1) Analysis of data on the links between sites is a perspective direction, but it requires a large and coherent training sample of web pages, the collection of which is a separate challenge. (2) When analyzing the responses of WhoIs servers we experienced difficulties caused by the complexity of messages analysis from different servers; it is a consequence of the lack of a common format of such answers. (3) Using existing lists of classified sites may have a place in the final classification scheme as an expert opinion, allowing to evaluate the result of the system, and as one of the sources of construction of provisional estimation of the site belonging to one category or another. In connection with the low quality of the original data in public databases, these lists were used only as part of the training and testing samples for the analysis of the quality of other classification techniques. (4) History of previous classifications should be accumulated over a sufficiently long period of time, so the research on evaluation of this information source continues.

5   Implementation

For downloading data from the sites, training classifiers and their testing we developed a software tool, called Web Classification Manager. The tool is developed in Java programming language and uses the following software libraries: Jsoup HTML parser 1.7.1 [9] and Rapid Miner 5.2 [25].

The software tool operates in the following three main modes:

1. Source data preparation mode. This mode is used for the preparation of the training and testing databases that do not depend on changes in the Internet. At the same time according to the prepared list of URL addresses of already classified web pages their HTML representation is loaded, from which, in turn, attributes for training of classifiers are selected. The list of classified web pages serves as the input data.

2. Training mode. This mode is designed for automated preparation of the trained model of the web page classifier. The features of classified web pages act as the input data. The result is a trained classifier model.

3. Testing mode. It allows assessing the quality of the trained model. Indicators for which a classifier is being evaluated are described in details above. The features of classified web pages and the trained classifier model serve as the input data. The result is a set of indicators characterizing the efficiency of the trained classifier model.

The tool provides the ability to perform in automated mode a wide variety of operations, and provides a "one-touch" tool for training and testing of the classifier models. During testing of the tool, we revealed that when the training set of web pages consists of 100 000 URL addresses preparation time and testing of the model takes about 70 hours (including 10-15 minutes of operator's work). Using the tool, we prepared: (1) the training and testing samples of web pages, as well as (2) the trained basic and combined classifiers based on the analysis of different kinds of information.

6   Experiments and Collected Statistics

6.1   Preparation of Training and Testing Samples

This stage can be divided into the following sub-steps: (1) creation of the list of categories of web pages; (2) preparation of the input lists of URL addresses of web pages; (3) loading the web content to the internal storage; (4) data pre-processing and extraction of features, which will be used to train classifier models.

Creation the list of categories of web pages. Quality of trained classifiers largely depends on the chosen categories of classification. For example, in some taxonomies, categorization of "hate" and "cruelty" are separated, however, there are many sites that cannot be clearly attributed to one or another category. Similar problems are "hacking", etc. Providing the list of categories it is necessary to maintain balance of detail and precision. Presence in the same list of categories such as "computers" and "software", "sport" and "fitness", etc. can cause reduction of classification quality.

Preparation and analysis of the initial lists of URL addresses of web pages. In this sub-step, the problems can be associated with using various sources of classified web pages for different categories. This occurs because the boundary between the categories is often subjective, leading to problems when training the classifier.

Loading the web content to the internal storage. Here the additional complexity is given by web pages with dynamic content (for example, web pages with Javascript). The content of such pages may depend on many factors and, as a consequence, the loaded page may not always match what the user can see through the browser.

Data pre-processing. The main objective here is to choose the most informative features that characterize the categories by which classification will be carried out. For example, category "news" is extremely difficult to determine on textual content, but the web pages of this category often have similar structure tags. Thus, the list of selected indicators directly depends on the selected categories.

As the *input data source* the Open Directory Project (ODP), also known as DMOZ was chosen. After several experiments it was found that it contains a lot of misclassified sites that decrease the further classification precision. On the last stages of experiments the manually tested site's list provided by F-Secure company was used.

6.2   Classification by Textual Information

To assess the quality of work of various classification techniques the cross-validation method was used (see Table 1).

Table 1. Comparison of quality of work of classifiers based on SVM, DT and NB

	Precision, %	Recall, %	Accuracy, %	F-measure, %
SVM	83.44	15,83	16,44	16,95
DT	82,34	17,66	17,72	20,75
NB	74,68	39,96	39,72	43,54

After comparison of quality of classification of text information using a variety of machine learning methods, the DT method was selected, as it showed a quite high level of precision with adequate results in recall, accuracy and F-measure.

Based on the results of the experiments carried out we can make the following conclusions: (1) On experiments, dealing with finding the optimal size of feature dictionary, we can conclude that the expansion of the dictionary improves the accuracy of classification only up to some limit, and after that information saturation leads to the fact that accuracy has been increasing slightly. (2) Each classifier determines with maximal accuracy different categories, hence the use of combination of individual classifiers of different types in different categories will lead to a significant increase in accuracy. (3) Introduction of a new category "unknown" and attributing to it of all web pages that cannot be exactly attributed, improved the accuracy of combined classifiers. Fig. 2 shows the analysis of different approaches to the choice of wordforms for constructing a dictionary (division into words based on analysis of non-alphabetic symbols, separation into sentences, separation into tokens, replacing words by hyponyms, replacing words by hyperonyms). The figure shows that the best accuracy was demonstrated by the method of partitioning the text on the basis of tokens.

6.3   Classification Based on URL Addresses

Experiments have shown that each of the categories has its specific well-defined set of n-grams, appearance of which in URL address is more likely.

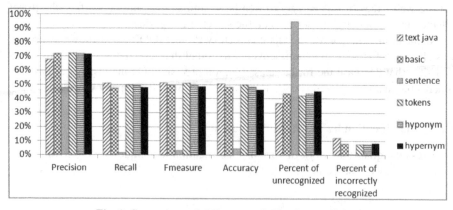

Fig. 2. Comparison of different methods of text splitting

However, in the general case, such n-grams are not necessarily present in the URL addresses of the corresponding categories. This explains the relatively high values of the precision of the target class at low values of the recall. This means that if a separate classifier trained on the attributes retrieved from the URL address, informs that the object belongs to its target category, then it can be trusted. However, due to the specificity of these data, there will be a set of objects not related to any category. The experimental results also showed that the technique of classification based on n-grams can be used in a complex solution that is based on multi-expert paradigm where each expert is responsible for decision making on attributing the classified URL address to the target category. From such a decision one may expect sufficiently high values of the precision, but it will be efficient for the 25-40% of the targets. Possible ways to enhance this approach is to use a more balanced and filled list of features and data.

6.4 Combined Classification

In general, the classifier based on the analysis of textual information shows the best results (high precision and recall) and the classifier based on URL addresses gives greater precision compared with the analysis of the structure of HTML. However, the combined use of individual classifiers based on these aspects of data that describe the object, gives significant improvement in the classification results.

Fig. 3 shows the results of the classification of the test data set using models trained on the text, models trained on the URL, the combined model trained on the text and tags, and the combined model trained on the text, tags and URL. This figure shows the values of indicators such as precision, recall, F-measure, accuracy, the percentage of "unknown" and misclassified web pages. The experimental results confirm the increase in classification performance when using the combined classification approach.

6.5 Classification of Web Pages in Languages Other Than English

To classify web pages in languages other than English, the first phases of preparation of input data of the testing sample were modified.

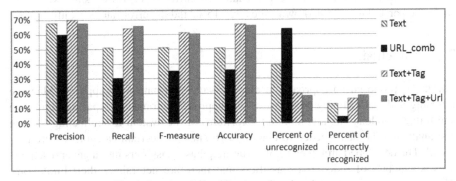

Fig. 3. Results of classification of testing data set

After loading the text content from web pages, the translation into English (which was used for the training of classifiers) was performed. Machine translation was performed using Yandex.Translate system [30], and the language of the original web page was determined automatically. Further stages of the web pages classification are similar to those performed in the classification of sites in English.

To verify the proposed approach the testing samples were formed for web pages belonging to the category "adult". For sites on the English, French and German the results are 94, 91 and 96%, respectively. These results show high classification quality of the "adult" category in German and French. However, it should be noted that the testing sample contains only the sites of these categories. In the case of inclusion in the sample of web pages belonging to other categories, F-measure is slightly reduced, as the classifier is likely to include web pages of other categories into the category of "adult" (type II errors).

It should be also noted that the quality of classification of web pages in German was slightly better than a similar indicator for English. This is mainly due the fact that some words having different connotations are translated as a single word, which reduces the diversity of wordforms, and as a result, can lead to better classification.

6.6 Comparison with Existing Analogues

For making comparison within the limits defined above, there were considered indicators of accuracy given in the papers of Chakrabarti et al. [4], Qi et al. [23], Calado et al. [3], Patil et al. [22] (75, 91, 81 and 89%, respectively) that have values that do not contradict principally to the values that we obtained (average accuracy across all available categories is about 67%). However, it should be noted that the result obtained of this paper was averaged. It does not display the accuracy achieved for certain categories (in the experiments we considered more than 20 categories, and, for example, for the categories "cigarettes", "dating", "adult", "casino" and "marijuana" the accuracy ranged from 87% up to 96%, but at the same time, the categories "cults", "occultism", "racism" and "religion" showed less than 50% accuracy), and it also ignores the importance of the additional "compensatory" mechanism counteracting the problem of false positives by using "unknown" as a category used to classify the "default" object (in this case the percentage of correctly classified sites was 67%, while the percentage of "unknown" sites was 15%, and the error rate was only 18%).

7   Conclusion

The paper proposed an approach to classify web information by applying Machine Learning and Data Mining techniques. The approach automates preparing input data and trained models. Architecture and algorithms of the system for collecting, storing and analyzing the data needed to classify web sites into certain categories were proposed. The developed architecture of combining base classifiers into a general scheme uses advantages of individual classifiers and therefore neutralizes their limitations. The software system to automate the classification of web pages was implemented.

Experiments to identify the main problems in the construction of web page classifiers were carried out. During experiments a lot of time was spent for understanding why certain classifier's decisions are wrong or good, testing different hypothesis and checking solutions. As a result, experiments showed high FP-proof classification accuracy for certain categories. This confirms feasibility of using the technology in systems of blocking websites with inappropriate content.

Further work on the development of the approach includes research areas aimed at improving the quality of decisions taken by the developed system. Very important goal of future work is taking context into account in text analysis.

Acknowledgements. This research is being supported by TEKES as part of the Data to Intelligence program of DIGILE (Finnish Strategic Centre for Science, Technology and Innovation in the field of ICT and digital business), the grants of the Russian Foundation of Basic Research (13-01-00843, 13-07-13159, 14-07-00697, 14-07-00417), the Program of fundamental research of the Department for Nanotechnologies and Informational Technologies of the Russian Academy of Sciences, the state project "Organization of scientific research" of the main part of the state plan of the Board of Education of Russia, and the SPbNRU ITMO project.

References

1. Agrwal, R., Srikant, R.: First algorithms for mining association rules. In: Proc. of the 20th Very Large Data Bases Conference, pp. 487–499 (1994)
2. Baykan, E., Henzinger, M., Marian, L., Weber, I.: Purely URL-based topic classification. In: Proc. of the WWW 2009, New York, USA, pp. 1109–1110 (2009)
3. Calado, P., Cristo, M., Moura, E., Ziviani, N., Ribeiro-Neto, B., Goncalves, M.A.: Combining link-based and content-based methods for web document classification. In: Proc. of the CIKM 2003, New York, USA, pp. 394–401 (2003)
4. Chakrabarti, S., Dom, B., Agrawal, R., Raghavan, P.: Scalable feature selection, classification and signature generation for organizing large text databases into hierarchical topic taxonomies. The Intern. Journ. on Very Large Data Bases 7(3), 163–178 (1998)
5. Cooley, R., Mobasher, B., Srivastava, J.: Web mining: Information and pattern discovery on the world wide web. In: Proc. of the ICTAI 1997, pp. 558–567 (1997)
6. Dumais, S., Chen, H.: Hierarchical classification of Web content. In: Proc. of the SIGIR 2000, pp. 256–263. ACM, New York (2000)
7. Dumais, S.T., Platt, J., Heckermann, D., Sahami, M.: Inductive learning algorithms and representations for text categorization. In: Proc. of the CIKM 1998, pp. 148–155 (1998)
8. F-Secure company, http://www.f-secure.com/
9. Java HTML Parser, http://jsoup.org/
10. Joachims, T.: Text categorization with support vector machines: learning with many relevant features. In: Nédellec, C., Rouveirol, C. (eds.) ECML 1998. LNCS, vol. 1398, pp. 137–142. Springer, Heidelberg (1998)
11. Kan, M.Y., Thi, H.O.N.: Fast webpage classification using url features. In: Proc. of the CIKM 2005, New York, USA, pp. 325–326 (2005)
12. Kan, M.Y.: Web page classification without the web page. In: Proc. of the WWW Alt. 2004, New York, USA, pp. 262–263 (2004)

13. Kleinberg, J.M., Kumar, R., Raghavan, P., Rajagopalan, S., Tomkins, A.S.: The Web as a Graph: Measurements, Models, and Methods. In: Asano, T., Imai, H., Lee, D.T., Nakano, S., Tokuyama, T. (eds.) COCOON 1999. LNCS, vol. 1627, pp. 1–17. Springer, Heidelberg (1999)
14. Komashinskiy, D.V., Kotenko, I.V., Chechulin, A.A.: Categorization of web sites for inadmissible web pages blocking. High Availability Systems (2), 102–106 (2011)
15. Kotenko, I.V., Chechulin, A.A., Shorov, A.V., Komashinkiy, D.V.: Automatic system for categorization of websites for blocking web pages with inappropriate. High Availability Systems (3), 119–127 (2013)
16. Kwon, O.W., Lee, J.H.: Text categorization based on k-nearest neighbor approach for web site classification. Information Processing and Management: an International Journal 29(1), 25–44 (2003)
17. Kwon, O.W., Lee, J.H.: Web page classification based on k-nearest neighbor approach. In: Proc. of the IRAL 2000, New York, USA, pp. 9–15 (2000)
18. Lai, Y.S., Wu, C.H.: Meaningful term extraction and discriminative term selection in text categorization via unknown-word methodology. ACM Transactions on Asian Language Information Processing (TALIP) 1(1), 34–64 (2002)
19. Lam, W., Ho, C.Y.: Using a generalized instance set for automatic text categorization. In: Proc. of the SIGIR 1998, Melbourne, Australia, pp. 81–89 (1998)
20. Lewis, D.D.: An evaluation of phrasal and clustered representations on a text categorization task. In: Proc. of the SIGIR 1992, Copenhagen, Denmark, pp. 37–50 (1992)
21. McCallum, A., Nigam, K.: A comparison of event models for naive Bayes text classification. In: Proc. of the AAAI/ICML 1998, pp. 41–48. AAAI Press (1998)
22. Patil, A., Pawar, B.: Automated Classification of Web Sites using Naive Bayessian Algorithm. In: Proc. of the IMECS 2012, vol. 1, p. 466 (2012)
23. Qi, X., Davison, B.D.: Knowing a Web Page by the Company It Keeps. In: Proc. of the CIKM 2006, pp. 228–237 (2006)
24. Qi, X., Davison, B.D.: Web Page Classification: Features and algorithms. ACM Computing Surveys (CSUR) 41(2), article No.12 (2009)
25. RapidMiner, http://rapid-i.com/content/view/181/190/
26. Schauble, P.: Multimedia Information Retrieval: Content-Based Information Retrieval from Large Text and Audio Databases. The Springer International Series in Engineering and Computer Science, pp. 49–59. Kluwer Academic Publishers, Norwell (1997)
27. Shibu, S., Vishwakarma, A., Bhargava, N.: A combination approach for Web Page Classification using Page Rank and Feature Selection Technique. International Journal of Computer Theory and Engineering 2(6), 897–900 (2010)
28. Tsukada, M., Washio, T., Motoda, H.: Automatic Web-Page Classification by Using Machine Learning Methods. In: Zhong, N., Yao, Y., Ohsuga, S., Liu, J. (eds.) WI 2001. LNCS (LNAI), vol. 2198, pp. 303–313. Springer, Heidelberg (2001)
29. Xu, Z., Yan, F., Qin, J., Zhu, H.: A Web Page Classification Algorithm Based on Link Information. In: Proc. of the DCABES 2011, pp. 82–86. IEEE Computer Society (2011)
30. Yandex. Translate API: http://api.yandex.com/translate/
31. Yang, Y., Liu, X.: A re-examination of text categorization methods. In: Proc. of the SIGIR 1999, Berkeley, CA, pp. 42–49 (1999)

Extending PubMed Related Article (PMRA) for Multiple Citations

Sachintha Pitigala[1] and Cen Li[2]

[1] Center for Computational Sciences, MTSU, Murfreesboro, TN, USA
[2] Department of Computer Science, MTSU, Murfreesboro, TN, USA
spp2k@mtmail.mtsu.edu, cen.li@mtsu.edu

Abstract. PubMed is the most comprehensive citation database in the field of biomedicine. It contains over 23 million citations from MEDLINE, life science journals and books. However, retrieving relevant information from PubMed is challenging due to its size and rapid growth. Keyword based information retrieval is not adequate in PubMed. Many tools have been developed to enhance the quality of information retrieval from PubMed. PubMed Related Article (PMRA) feature is one approach developed to help the users retrieve information efficiently. It finds highly related citations to a given citation. This study focuses on extending the PMRA feature to multiple citations in the context of personalized information retrieval. Our experimental results show that the extended PMRA feature using the words appearing in two or more citations is able to find more relevant articles than using the PMRA feature on individual PubMed citations.

Keywords: PubMed, Information Retrieval, Similarity Measures, PubMed Related Citations, Personalized Article Retrieval System.

1 Introduction

National Library of Medicine (NLM) started to index biomedical and life science journal articles in 1960's. The indexed citations were kept in the Medline citation database. Currently, NLM provides access to over 19 million citations dating back to 1946 [1]. In 1996, National Center for Biotechnology Information (NCBI) at NLM introduced PubMed citation database. PubMed provides access to over 23 million citations in the field of biomedicine [2]. Primarily, it allows free access to Medline citation database via internet. PubMed contains more citations than the Medline database covering the in-progress Medline citations, out of scope citations, "Ahead of Print" citations, and NCBI bookshelf citations. Therefore, PubMed is the most comprehensive citation database in the field of biomedicine.

Typical users of PubMed search for relevant articles to their specific research interests by entering one or more query terms on PubMeds web interface. This task has become more and more challenging due to PubMeds rapid growth of citations. Often times, too many citations were returned as a result of the query,

P. Perner (Ed.): ICDM 2014, LNAI 8557, pp. 55–69, 2014.

while many of the returned citations are not directly relevant to the information need. To improve the quality of retrieval from PubMed, NCBI and other academic and industry groups have developed many tools. Two main approaches have been used to enhance the information retrieval systems. The first approach builds supplementary tools for the original PubMed search interface. For example, PubMed advanced search feature ([3], [4]), PubMed auto query suggestions [5], PubMed automatic term mapping [6], and PubMed related article feature [7] are some of the PubMed supplementary tools to enhanced information retrieval from PubMed. The second approach builds entirely new tools which are complementary to the PubMed search interface. MedlineRanker [8], MScanner [9], PubFinder [10], Caipirini [11] and Hakia [12] are examples in this category.

Some popular approaches in building complementary tools are based on text classification methods ([8], [9], [10]), semantic based methods ([12], [13]) and special input (set of genes or set of protein names) based methods ([11], [14]). This study focuses on developing a complementary search tool to PubMed using text classification approach. Two recent PubMed complementary tools are MedlineRanker [8] and MScanner [9]. Each system starts with a set of abstracts that are known to be relevant to a query topic of interest, or information need. Then, it trains a Naive Bayes text classifier based on these abstracts. The learned text classifier is then used to find the relevant documents to the information need from the PubMed. However, in order to get good results from MedlineRanker and MScanner, at least 100 highly relevant abstracts need to be provided by the user ([8], [9]). This is a requirement that is not easily satisfiable by most users. Finding hundreds of abstracts that have been confirmed to be relevant to certain information need is a time consuming process.

It is desirable to have a document retrieval system that allows one to retrieve articles pertinent to his study, only requiring a small set of abstracts confirmed to be relevant to the information need. The ultimate goal of this study is to develop a complementary tool to PubMed, such that when given an information need, it is capable of training a text classifier using a small number of PubMed abstracts and retrieving highly relevant articles.

It is well known to the text mining community, it is difficult to train text classifier with good accuracy based on small data set. Therefore, an important step of building the proposed system is to identify a proper technique to increase the training set size, based on the small set of abstracts provided by the user. One approach to this problem is based on interactive user inputs. It asks explicit user feedback about the relevance of the articles extracted solely based on the small data set. The PubMed abstracts deemed relevant by the user are added to the training set. This process is repeated until the system gets sufficient amount of relevant articles. The search tool RefMed [15] was developed based on this multi-level relevance feedback with the RankSVM learning method. This multi-level relevance feedback method allows the user to express the user information need more thoroughly. However, this approach is also time consuming and tiresome which requires proficient knowledge about the biomedicine field. In addition, users need to be cautious about the feedback inputs. Since, less relevant abstracts admitted into the data set may decrease the accuracy of the classifier learned. After multiple iterations of inaccurate learning, the classifier may produce final results far from the initial information need.

Another approach is to increase the training set size by finding the most similar abstracts to the input seed abstracts based on document similarity. This approach is more efficient than the first approach, and it does not require user feedbacks. However, most of the standard similarity measures such as Pearson Correlation Coefficient [16], Cosine Similarity [17] are too general and not suitable for finding similar document from large databases such as PubMed. What we need is a similarity measure that can be used to find documents similar to the seed abstracts from a large database.

This paper focuses on extending the PubMed Related Article (PMRA) [7] measure for finding similar articles for multiple citations. PMRA is a well-established tool in PubMed. It is capable of finding relevant articles from the entire PubMed database for a given PubMed citation. PubMed real world log analysis shows that roughly a fifth of all non-trivial PubMed sessions used the related article feature [18]. The questions we would like to answer are:

- *"Can we find the articles similar to the set of seed abstracts by simply combining the individual PubMed related article lists of individual seed abstract?"*;
- *"Is it more accurate to combine the seed abstracts into one single super-citation and find articles that are similar to this super-citation?"*; and
- *"What is the best way to extend the PMRA method when there are multiple seed abstracts presented?"*.

In order to answer these questions we propose a number of extension approaches to PMRA, and a series of experiments that compare these approaches using the TREC 2005 genomic track data [19].

The rest of the paper is organized as the following: Section 2 discusses similarity measures used for text classification and the theory behind the original PMRA method. Section 3 presents the proposed extended PMRA methods, Dataset, Preprocessing steps and Procedure of estimating parameters. Section 4 describes the experimental results of the proposed methods on the TREC 2005 data. Section 5 draws the conclusions about the study and presents the future research directions.

2   Background

A similarity measure gives a formal definition to quantify the similarity between two instances. A distance measure quantifies how far apart two instances are. Some of the popular distance measures are Euclidean distance [20], City Block (Manhattan) distance [21] and Chebyshev distance [22]. Both similarity and distance measures are widely used in information retrieval, clustering algorithms and many other data mining applications.Distance/similarity measures can be divided into two broader categories as vector based and probabilistic based measures. Pearson Correlation Coefficient [16], Cosine Similarity [17], Jaccard Coefficient [23] and Tanimoto Coefficient [24] are some of the vector based similarity measures. Fidelity similarity (Bhattacharyya coefficient or Hellinger affinity) [25], PMRA method [7], bm25 ([26],[27]) are examples of probabilistic based similarity measures.

PMRA probabilistic similarity measure was used to develop the related article feature in PubMed. It finds articles similar to a chosen PubMed citation from the entire PubMed database. When a user selects a citation from the PubMed search results, the right panel of the browser window displays citations that have the highest PMRA similarity value, e.g., the closest matching, to the chosen citation. The list of the most similar citations forms the related citation list. The related citation list for each article in PubMed is pre-calculated, and pre-sorted according the PMRA value [7]. The calculation and sorting of PMRA lists are done at the back-end and PubMed is updated periodically with the new PMRA scores.

2.1   PMRA Method

Given that document d is deemed related to one's information need, PMRA computes the relatedness of document c in terms of the posterior probability $P(c|d)$, where c can be any document in PubMed. Assuming a document can be decomposed into a set of mutually exclusive and exhaustive "topics" $s_1, s_2, ., s_N$. $P(c|d)$ can be computed as following equation 1:

$$P(c|d) = \sum_{j=1}^{N} P(c|s_j)P(s_j|d) \tag{1}$$

Expanding $P(s_j|d)$ using the Bayes theorem, we obtained equation 2.

$$P(c|d) = \frac{\sum_{j=1}^{N} P(c|s_j)P(d|s_j)P(s_j)}{\sum_{j=1}^{N} P(d|s_j)P(s_j)} \tag{2}$$

For a user selected document d, the denominator of equation 2 remains constant for any document c. Therefore, the denominator of equation 2 can be ignored and the following criteria can be used to rank documents based on their relatedness/similarity.

$$P(c|d) \propto \sum_{j=1}^{N} P(c|s_j)P(d|s_j)P(s_j) \tag{3}$$

Here $P(c|s_j)$ is the probability that the user find an interest in document c, given an interest in topic s_j. Similarly, $P(d|s_j)$ is the probability that the user find an interest in document d, given an interest in topic s_j. $P(s_j)$ is the prior probability of the topic s_j i.e., the fraction of all documents that discusses the topic s_j. Therefore, relevance of a document c to the given document d can be computed by summing up the product of $P(c|s_j), P(d|s_j)$ and $P(s_j)$ across all the topics [7].

In order to estimate $P(c|s_j), P(d|s_j)$ and $P(s_j)$, PMRA introduced a concept called *eliteness* [7]. *Eliteness* explains whether a given document d is about a particular topic s_j or not. The original PMRA method assumes that each word in the PubMed citation (title, abstract and MeSH term list) represents a topic (s_j). Moreover, each word (term) in the PubMed citation represents an idea or concept in the document. A term t_i is elite for document d, if it represents the topic s_j. Otherwise, term t_i is non-elite for document d. Equation 4 can be derived using the *eliteness* concept and Bayes theorem [7]. Let, E represent the *eliteness* of a term in document d, and $\bar{E}$ represent the *non-eliteness* of a term in document d. The probability a term is *elite* in a document is conditioned on the number of times, k, that term appears in the document:

$$P(E|k) = \frac{P(k|E)P(E)}{P(k|E)P(E) + P(k|\bar{E})P(\bar{E})} = \left(1 + \frac{P(k|\bar{E})P(\bar{E})}{P(k|E)P(E)}\right)^{-1} \quad (4)$$

$P(k|E)$ and $P(k|\bar{E})$ are calculated using Poisson distributions as shown in Equations 5 and 6.

$$P(k|E) = \frac{\lambda^k e^{-\lambda}}{k!} \quad (5)$$

$$P(k|\bar{E}) = \frac{\mu^k e^{-\mu}}{k!} \quad (6)$$

where λ is the mean of the Poisson distribution of the *elite* case for the given term, and μ is the mean of the Poisson distribution of the *non-elite* case for the given term. First, substitute equation 5 and 6 values in to equation 4. Then, applying document length normalization and algebraic manipulations to equation 4, we derived equation 7.

$$P(E|k) = \left(1 + \frac{\mu^k e^{-\mu} P(\bar{E})}{\lambda^k e^{-\lambda} P(E)}\right)^{-1} = \left[1 + \eta \left(\frac{\mu}{\lambda}\right)^k e^{-(\mu-\lambda)l}\right]^{-1} \quad (7)$$

where l is the length of the document and $\eta = P(\bar{E})/P(E)$.

Then, we combine the concept of *eliteness* with the relatedness concept of two documents. $P(E|k)$ is used to estimate $P(c|s_j)$ and $P(d|s_j)$ in the $P(c|d)$ model. To efficiently calculate the similarity values, the $P(s_j)$ is estimated using the inverse document frequency of term (topic) t_j, idf_{t_j}. Then, the following weighting function and the similarity function are derived to calculate the similarity of the two documents.

$$P(c|d) \propto sim(c,d) = \sum_{j=1}^{N} [P(E|k)]_{t_j,c} \cdot [P(E|k)]_{t_j,d} \cdot idf_{t_j} \quad (8)$$

$$sim(c,d) = \sum_{j=1}^{N} \left[1 + \eta \left(\frac{\mu}{\lambda}\right)^k e^{-(\mu-\lambda)l}\right]_{t_j,c}^{-1} \cdot \sqrt{idf_{t_j}} \cdot \left[1 + \eta \left(\frac{\mu}{\lambda}\right)^k e^{-(\mu-\lambda)l}\right]_{t_j,d}^{-1} \cdot \sqrt{idf_{t_j}} \quad (9)$$

$$w_t = \left[1 + \eta \left(\frac{\mu}{\lambda}\right)^k e^{-(\mu-\lambda)l}\right]^{-1} \cdot \sqrt{idf_t} \quad (10)$$

$$sim(c, d) = \sum_{j=1}^{N} w_{t_j,c} \cdot w_{t_j,d} \tag{11}$$

where w_t calculates the term weight for a given document. Similarity between the two documents is computed with an inner product of the term weights as in Equation 11.

2.2 Parameter Estimation in PMRA

PMRA similarity calculation requires that a number of parameters, λ, μ, η, be estimated. A simplifying assumption has been made for the *elite* and *non-elite* Poisson distributions: half of the terms in the document are *elite* and the other half of the terms are *non-elite*. This assumption leads to equation 12, a model similar to the maximum entropy models used in natural language processing ([7],[28]).

$$\eta\left(\frac{\mu}{\lambda}\right) = \frac{P(\bar{E})\mu}{P(E)\lambda} = 1 \tag{12}$$

The weighting scheme expressed in the equation 10 can then be re-written as:

$$w_t = \left[1 + \left(\frac{\mu}{\lambda}\right)^{k-1} e^{-(\mu-\lambda)l}\right]^{-1} \cdot \sqrt{idf_t} \tag{13}$$

This way, PMRA reduces the number of parameters to be estimated from three to two. Medical Subject Heading (MeSH) information in Medline was used to estimate λ and μ. MeSH descriptors to each PubMed indexed citation are assigned manually by experts in the field of biomedicine. Therefore, terms in the MeSH descriptors can be considered as *elite* terms for the citations. The terms in the citation that do not appear in the MeSH descriptors are considered *non-elite* terms for the citation. The average appearance of a given *elite* term (λ) or a given *non-elite* term (μ) can be calculated based on a collection of PubMed citations.

The following section explains the methodology of this study. In particularly, it explains how PMRA is extended for multiple citations, the dataset used for this experiment, and the data pre-processing steps.

3 Methodology

As stated in Section 1, the ultimate goal of this project is to develop an enhanced information retrieval system for PubMed that can suggest related articles based on classifier learned from small number of user-defined citations. This paper discusses our approaches to increase the training set size based on the small set of citations provided by the user. PMRA measure is extended for this purpose. We experimentally evaluate these approaches using the TREC 2005 genomic track data [19].

3.1 Extending the PMRA Similarity Measure

PMRA was developed to find the relevant citations for a single user selected citation. Currently, PMRA method is not directly applicable for finding the relevant citations for multiple user selected citations. Next we discuss a number of approaches to extend PMRA for multiple citations.

The straightforward way of extending PMRA for multiple citations is to combine the PubMed related article lists obtained from the individual seed citations, and sort all the derived articles according to their PMRA similarity values. We refer to this method as the Basic method. The PMRA related article list for individual citations is pre-calculated in PubMed. Therefore, the Basic method can be completed in a very short time. However, this method is not good at capturing the overall user concept or idea of the information need expressed through multiple citations.

The second approach is to combine multiple citations into a single citation and to find the relevant citations to this newly formed citation. This method is slower than the first approach because the newly formed citation is not present in the PubMed database, therefore no pre-computed list is available. But, this approach gives a better representation of the particular user information need by taking into account information present in all the user-defined citations.

There are multiple ways of combing the set of seed articles into a single citation:

- The first method, the All-inclusive method, simply combines the terms from all the seed citations;
- The second method, the Intersection method, forms the new citation by only including terms that simultaneously appeared in every single seed citation, i.e., intersection of all seed citations;
- The third method, the At-least-two method, forms the new citation by including terms appearing in at least two seed citations.

We experimentally compare the effectiveness of these four methods: the Basic method, the All-inclusive method, the Intersection method, and the At-least-two method in the Section 4. Next, the data used for the experiments is discussed.

3.2 TREC 2005 Dataset

A subset of TREC 2005 genomic track [19] was used in this study. In particularly, Ad-Hoc retrieval task dataset from the TREC 2005 genomic track was used. This is the same dataset used in the original PMRA experiment study [7]. It contains 50 different information needs (topics) from biologists. The entire document collection for the 50 topics contains 34,633 unique PubMed citations. Each topic corresponds to a different subset of documents ranging in size from 290 to 1356 documents. Relevance of each document to the given topic was judged by a group of scientists. According to their opinion all the documents in the document pool were labeled as: Definitely Relevant (DR), Possibly Relevant (PR) or Non Relevant (NR). The ten topics having the highest number of relevant documents (definitely relevant and possibly relevant) were used in this study. Table 1 shows the document distribution of those ten topics.

Table 1. Ten topics (information needs) that contain the highest number of relevant documents in the TREC 2005 genomic track dataset

Topic ID	# Definitely Relevant documents	# Possibly Relevant documents	# Non Relevant documents	# Total documents
117	527	182	385	1094
146	370	67	388	825
120	223	122	182	527
114	210	169	375	754
126	190	117	1013	1320
109	165	14	210	389
142	151	120	257	528
111	109	93	473	675
107	76	114	294	484
108	76	127	889	1092

3.3 Data Preprocessing

The TREC 2005 dataset has a list of PMID's for all the topics along with their relevance judgment to the given topic. First, all the PubMed citations for the given 50 topics were downloaded from the PubMed using the Entrenz utilities provided by the NCBI [29]. When downloaded, the PubMed citations are in the XML format. First, PubMed citation title, abstract and the MeSH terms were extracted from the XML documents. All the other information such as details about the author, affiliation data and journal information were ignored in this study. Then, the title, abstract text and the MeSH terms were tokenized into list of terms. From the citation term list, stopwords [30] and words containing only digits were removed. Next, stemming was applied to obtain a normalized term list for the citation. Finally, the normalized terms from the title and MeSH (Medical Subject Headings) terms with subheading qualifier were added again to the normalized term list to give more weight to those terms. Term list for each of the 34,633 citations was constructed using this data pre-processing procedure. These lists were used to estimate λ and μ parameters and calculate the similarities between citations.

3.4 Estimating λ, μ Parameters

To estimate λ and μ parameters, the normalized term list for a given article was divided into two sets, i.e. *elite* terms and *non-elite* terms. From the normalized term list, terms appearing only in the MeSH terms were labeled as *elite* terms for the given citation. Next, all the terms not appearing in the *elite* term list were labeled as *non-elite* terms for the given citation. This process was repeated for the entire collection of 34,633 documents to obtain the *elite* and *non-elite* term lists for each citation. An *elite* word dictionary was then created with the unique *elite* terms along with their average term frequencies. The average term frequency for a given *elite* term was calculated using equation 14.

$$at f_t = \frac{\sum_{i=1}^{N} t f_{t,d_i}}{df_t} \tag{14}$$

where, $at f_t$ is the average term frequency for the given *elite* term, $t f_{t,d_i}$ is the term frequency (number of occurrences) for the term t in the ith document's *elite* term list, N is the total number of documents in the collection and df_t is the total number of documents which has the term t in it's elite term list.

The average term frequency defined in equation 14 corresponds to the Poisson mean (λ) for a given *elite* term. Similarly, the average term frequencies for the *non-elite* terms can be calculated using the *non-elite* term lists in the document collection. These *non-elite* average term frequencies corresponds to the Poisson mean (μ) to the given *non-elite* terms.

3.5 Experiment Procedure

For our experiments, the Definitely Relevant and Possibly Relevant citations for a given topic (information need) were combined and labeled as relevant citations. Next, n citations were randomly selected from the relevant citation set and labeled as user seeds. In this study, the number of user seed citations (n) was varied from 1 to 10 according to the experiments.

To compare the effectiveness of the four PMRA extension methods, each method is applied to derive a different related citation list based on n user selected seed citations. The citations included in each resulting related citation list are then sorted in descending order based on their PMRA values. The precision of the method is computed in terms of the percentage of the top citations in the sorted list that were originally labeled as Definitely Relevant or Possibly Relevant, as shown in Equation 15.

$$Precision = \frac{\# of\, relevant\, citations}{\# of\, retrieved\, citations} \tag{15}$$

In this study, precision of the top five citations (P5), the top ten citations (P10), the top 20 citations (P20), the top 50 citations (P50) and the top 100 citations (P100) were calculated for each method. Each experiment was repeated ten times by randomly selecting different seed citations from the relevant set. The following section presents the experiment results obtained.

4 Results and Discussion

We experimentally compare the effectiveness of the four methods: the Basic method, the All-inclusive method, the Intersection method, and the At-least-two method, in finding related citations for a given set of seed citations. Ten experiments were conducted for each information need (topic) by changing the initial seed set size n from 1 to 10. Each experiment was repeated 10 times with different random seeds. Table 2 compares the overall average P5 precision of the four methods for each of the 10 information needs.

Table 2. Average P5 precision of the four methods for the ten information needs. The average was calculated over ten different seed set sizes. The boldface values indicate the highest P5 result for the given information need.

Topic ID	Basic method	All-inclusive method	Intersection method	At-least-two method
117	0.59	0.56	**0.74**	0.73
146	0.71	0.75	**0.86**	**0.86**
120	0.66	0.64	**0.93**	0.87
114	0.48	0.38	0.60	**0.67**
126	**0.27**	0.16	**0.27**	0.26
109	**0.95**	0.87	0.89	0.94
142	**0.77**	0.68	0.63	0.71
111	**0.59**	0.57	0.53	0.58
107	0.67	0.58	0.25	**0.68**
108	0.28	0.25	0.28	**0.48**

Results in Table 2 show that the Basic method produced comparable results to the results from the original PMRA study [7]. This makes the Basic method a good baseline to compare the other three methods. Results from this experiment also show that overall, the At-least-two method produced more relevant documents within the first five documents in the final output. For the 10 information needs, it has the highest P5 value, or when it isn't the highest, its P5 values are very close to the highest value. This is because the At-least-two method captures important words from different seeds and produce more specific combined citation for given information need.

The results also show that the All-inclusive method is not an effective method for combining multiple citations. It forms the new citation by taking into it all the words from all the seed citations. This causes the length, l, of the new citation to become much greater than that of a regular citation. Higher l value in equation 13 reduces the weighted values of the words in the combined citation. Therefore, this method produced less accurate results compared to other three methods.

Next, precision of the top 50 citations (P50) and precision of the top 100 citations (P100) were used test the effectiveness of four methods. Tables 3 and 4 present the average P50 and P100 precision for the 10 information needs respectively.

Tables 3 and 4 show the Intersection method and At-least-two method outperform the Basic and the All-inclusive methods. In fact, the Intersection method and At-least two method were able to produce related citation lists that are 20% more accurate than those produced by the other two methods for a number of topics. Also, the All-inclusive method produced the worst results using P50 and P100 precision measures.

Table 3. Average P50 measure of ten information needs. The average was calculated using ten different seed set sizes. The boldface values show the highest P50 result for the given information need.

Topic ID	Basic method	All-inclusive method	Intersection method	At-least-two method
117	0.55	0.49	0.68	**0.72**
146	0.58	0.56	**0.76**	0.75
120	0.60	0.51	**0.84**	0.78
114	0.38	0.31	0.49	**0.55**
126	0.21	0.19	**0.28**	0.26
109	0.83	0.71	0.78	**0.87**
142	**0.59**	0.53	0.57	0.54
111	0.45	0.41	0.40	**0.46**
107	0.50	0.44	0.50	**0.53**
108	0.36	0.30	0.21	**0.41**

Table 4. Average P100 measure of ten information needs. The average was calculated using ten different seed set sizes. The boldface values show the highest P100 result for the given information need.

Topic ID	Basic method	All-inclusive method	Intersection method	At-least-two method
117	0.53	0.47	**0.72**	0.70
146	0.51	0.48	**0.74**	0.69
120	0.55	0.45	**0.72**	**0.72**
114	0.34	0.28	0.45	**0.50**
126	0.20	0.17	**0.29**	0.25
109	0.74	0.58	0.74	**0.78**
142	0.48	0.42	**0.54**	0.46
111	0.40	0.35	0.34	**0.42**
107	0.42	0.37	**0.49**	0.45
108	0.31	0.25	0.17	**0.33**

From the results shown in Tables 3 and 4, the Intersection method and At-least-two method appear comparable across the 10 information needs. To determine which of these two methods is better, a statistical analysis is performed. The overall average and the 95% confidence interval for each method were calculated using all the experiments conducted in this study. 10 experiments were performed for each information need by changing seed set size from 1 to 10. For each seed set size, average of P5, P10, P20, P50, and P100 values were recorded. Therefore, 50 average measurements were calculated for each information need. Table 5 shows the average and the confidence interval for each of the four methods.

Table 5. The overall average and its 95% confidence interval of each method in this study

	All-inclusive method	Basic method	Intersection method	At-least-two method
Average	0.48	0.54	0.57	0.62
95% CI	(0.463, 0.497)	(0.523, 0.557)	(0.549, 0.591)	(0.602, 0.638)

Results in Table 5 clearly show that the At-least-two method produced the best results in this study. Its 95% confidence interval is higher than that from the other three methods. The All-inclusive method performed poorly in this study. The average accuracy of the Basic method and the Intersection method are quite close. The confidence intervals of these two methods overlapped. This indicates that the performance of the Basic method and the Intersection method are mostly similar.

To take a closer look at the performance of the At-least-two method, the average accuracies are plotted for different information needs across different initial seed set sizes. In Figure 1 and 2, accuracy was calculated using the average P5, P10, P20, P50 and P100 values for a given seed set size. Figure 1 shows the accuracy change over the initial seed set size for the first five information needs, and Figure 2 shows the accuracies over different seed set sizes for the next five information needs. It is observed that, for the majority of the information needs, the optimal performance is achieved when the seed set size is between 2 and 4. Afterwards, the accuracy values level off.

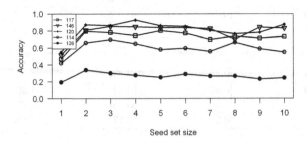

Fig. 1. Average accuracies for the first five information needs (117, 146, 120, 114, and 126) over different seed set sizes. Each data point is calculated using the P5, P10, P20, P50 and P100 measures for the given seed set size, and each P measure was calculated using 10 different random experiments.

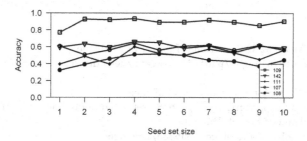

Fig. 2. Average accuracies for the next five information needs (109, 142, 111, 107 and 108) over different seed set sizes. Each data point is calculated using the similar procedure in figure 1.

5   Conclusions

This paper discusses the need to extend the PMRA similarity measure to work with multiple citations, and presents four ways to extend the original PMRA measure. To study the effectiveness of these methods, comparative analysis was conducted using the ten information needs in the TREC 2005 genomic track dataset [19].

Accuracy of the first five documents (P5) in the Basic method was comparable to the results given in the original PMRA study [7]. The At-least-two method is the best method to extend PMRA for multiple citations. This method best captures the important terms and discards the less important terms from the seed documents. In contrast, the Intersection method uses the terms appear in all seed documents, leading to a much smaller combined citation. The small number of terms from the combined citation is insufficient in accurately capturing similarity between citations. The performance of this method is comparable to that of the Basic method in this study. Overall, the All-inclusive method produced the least accurate results. The combined citation from this method is very long and contains a lot of less frequent terms. These less frequent terms and the length of the citation contributed to the low accuracy in finding the related citations.

The At-least two method generally achieved its maximum accuracy with only 4 seed citations. Adding more seed citations doesn't help to increase the accuracy under this experiment setting. The seed documents for a given experiment were randomly selected from the definitely relevant and possibly relevant documents. Therefore, a possibly relevant document which is less relevant to the information need can be selected to the seed set. This document will diverge the information need. Because of this, final relevant citation list can be less accurate. But, in practical situations, user can avoid this by selecting only relevant articles to the seed set. Future study can be conducted to reduce noise in the seed documents.

References

1. Fact Sheet-Medline, U.S. National Library of Medicine: `http://www.nlm.nih.gov/pubs/factsheets/medline.html` (retrieved August 25, 2013)
2. Pubmed, U.S. National Library of Medicine: `http://www.ncbi.nlm.nih.gov/pubmed` (retrieved September 20, 2013)
3. PubMed Advanced Search Builder, `http://www.ncbi.nlm.nih.gov/pubmed/advanced`
4. Chapma, D.: Advanced Search Features of PubMed. J. Can. Acad. Child Adolesc. Psychiatry 18(1), 58–59 (2009)
5. Lu, Z., Wilbur, W.J., McEntyre, J.R., Iskhakov, A., Szilagyi, L.: Finding Query Suggestions for PubMed. In: AMIA Annu. Symp. Proc. 2009, pp. 396–400 (2009) (published online November 14, 2009)
6. PubMed's Automatic Term Mapping Enhanced. NLM Tech Bulletin (341), e7 (November-December 2004)
7. Lin, J., Wilbur, W.: Pubmed related articles: a probabilistic topic-based model for content similarity. BMC Bioinformatics 8, 423 (2007)
8. Fontaine, J.F., Barbosa-Silva, A., Schaefer, M., et al.: MedlineRanker: flexible ranking of bio-medical literature. Nucleic Acids Res. 37, W141–W146 (2009)
9. Poulter, G., Rubin, D., et al.: MScanner: a classifier for retrieving Medline citations. BMC Bioinformatics 9, 108 (2008)
10. Goetz, T., Von Der Lieth, C.-W.: PubFinder: a tool for improving retrieval rate of relevant PubMed abstracts. Nucleic Acids Res. 33, W774 (2005)
11. Caipirini: using gene sets to rank literature. BioData Min. 5(1), 1 (2012)
12. Hakia (2008), `http://pubmed.hakia.com/` (date last accessed September 28, 2013)
13. Wang, J., Cetindil, I., Ji, S., et al.: Interactive and fuzzy search: a dynamic way to explore MEDLINE. Bioinformatics
14. Quertle (2009), `http://www.quertle.info` (date last accessed September 28, 2013)
15. Yu, H., Kim, T., Oh, J., et al.: Enabling multi-level relevance feedback on PubMed by integrating rank learning into DBMS. BMC Bioinformatics (2010)
16. Pearson, K.: Contributions to the mathematical theory of evolution, III, Regression, heredity, and panmixia. Philos. Trans. R. Soc. Lond. Ser. A 187, 253–318 (1896)
17. Manning, C.D., Raghavan, P., Schütze, H.: Introduction to Information Retrieval, pp. 120–122. Cambridge University Press (2008)
18. Lin, J., DiCuccio, M., Grigoryan, V., Wilbur, W.J.: Exploring the Effectiveness of Related Article Search in PubMed. Tech. Rep. LAMP-TR-145/CS-TR-4877/UMIACS-TR-2007-36/HCIL-2007-10. University of Maryland, College Park, Maryland (2007)
19. Hersh, W.R., Cohen, A.M., et al.: The Fourteenth Text Retrieval Conference (TREC 2005) NIST. TREC 2005 Genomics track overview (2005)
20. Duda, R.O., Hart, P.E., Stork, D.G.: Pattern Classification, 2nd edn., p. 187. Wiley-Interscience, New York (2001)
21. Krause, E.F.: Taxicab Geometry: An Adventure in Non-Euclidean Geometry
22. Cantrell, C.D.: Modern Mathematical Methods for Physicists and Engineers. Cambridge University Press (2000)
23. Tan, P.-N., Steinbach, M., Kumar, V.: Introduction to Data Mining (2005)
24. Theodoridis, S., Koutroumbas, K.: Pattern Recognition, 4th edn., p. 602, 605, 606. Academic Press, New York (2009)

25. Cha, S.: Comprehensive Survey on Distance/Similarity Measures between Probability Density Functions. International Journal of Mathematical Models and Methods in Applied Sciences 1(4), 300–307 (2007)
26. Robertson, S.E., Walker, S., Jones, S., Hancock-Beaulieu, M., Gatford, M.: Okapi at TREC-3. In: Proceedings of the 3rd Text REtrieval Conference, TREC-3 (1994)
27. Sparck Jones, K., Walker, S., Robertson, S.E.: A Probabilistic Model of Information Retrieval: Development and Comparative Experiments (Parts 1 and 2). Information Processing and Management 36(6), 779–840 (2000)
28. Berger, A., Pietra, S.D., Pietra, V.D.: A Maximum Entropy Approach to Natural Language Processing. Computational Linguistics 22, 39–71 (1996)
29. NCBI. Entrez Programming Utilities Help (2010),
    http://www.ncbi.nlm.nih.gov/books/NBK25501/
30. PubMed Stopwords, http://www.nlm.nih.gov/bsd/disted/pubmedtutorial/020_170.html (date last accessed August 24, 2013)

Variability of Behaviour in Electricity Load Profile Clustering; Who Does Things at the Same Time Each Day?

Ian Dent[1], Tony Craig[2], Uwe Aickelin[1], and Tom Rodden[1]

[1] School of Computer Science, University of Nottingham, Nottingham NG8 1BB, UK
psxid@nottingham.ac.uk
http://ima.ac.uk/dent
[2] The James Hutton Institute, Aberdeen, UK

Abstract. UK electricity market changes provide opportunities to alter households' electricity usage patterns for the benefit of the overall electricity network. Work on clustering similar households has concentrated on daily load profiles and the variability in regular household behaviours has not been considered. Those households with most variability in regular activities may be the most receptive to incentives to change timing. Whether using the variability of regular behaviour allows the creation of more consistent groupings of households is investigated and compared with daily load profile clustering. 204 UK households are analysed to find repeating patterns (motifs). Variability in the time of the motif is used as the basis for clustering households. Different clustering algorithms are assessed by the consistency of the results.

Findings show that variability of behaviour, using motifs, provides more consistent groupings of households across different clustering algorithms and allows for more efficient targeting of behaviour change interventions.

1 Background and Motivation

The electricity market in the UK is undergoing dramatic changes. Legal, social and political drivers for a more carbon efficient electricity network, along with the dramatically increased flow of data from households through the deployment of smart meters, requires a transformation of existing practices. In particular, the change of the frequency of sampling of electricity usage, by using smart meters, alters the level of understanding of households' behaviour that is possible [1].

One approach to address the pressures on the electricity network is the application of Demand Side Management (DSM) techniques to achieve changes in consumer behaviour. DSM is defined as "systematic utility and government activities designed to change the amount and/or timing of the customer's use of electricity" for the collective benefit of society, the utility company, and its customers [2]. The peak time for electricity usage in the UK is during the early evening and the successful application of techniques to reduce, or move, the peak usage would improve the overall efficiency of the electricity network.

P. Perner (Ed.): ICDM 2014, LNAI 8557, pp. 70–84, 2014.

To allow selection of appropriate DSM interventions, a good understanding of the existing behaviour of households is needed. Firstly, knowledge is needed on an individual household that can be deduced from house-wide electricity metering. Secondly, a method is required to group large numbers of households into a manageable number of archetypal groups where the members display similar characteristics. This approach allows for cost effective targeting of the most appropriate subset of customers whilst allowing the company management to deal with a manageable number of archetypes [3].

There is an extensive body of work on clustering households which includes comparing or combining timed meter readings to create additional attributes that contribute to the quality of the clustering [4]. However, little work has focused on how the daily activity patterns of the household vary from day to day and how this can be used for clustering. For instance, some households will be creatures of habit and will eat their evening meal at almost the same time each evening, whilst others have a much more variable activity pattern and will eat at different times. Ellegård and Palm [5] have investigated the variability of behaviour using diaries and interviews but have not used analysis of meter data.

Clustering households using their degree of variability in behaviour, as shown by electricity consumption, provides a way of identifying the subset of electricity users who may be most receptive to an intervention to influence their activity patterns. The intervention may be to reward households for NOT changing their current pattern of usage if it is already as desired by the utility company.

This paper addresses the question of whether making use of the variability of behaviour (as shown by the electricity meter data) provides "better" groupings of households for the purpose of DSM than those provided by using daily load profiles. The judgement of "better" is measured by implementing a number of different clustering techniques and measuring the degree of overlap between the clusters found. A consistent set of clusters across the different clustering algorithms implies a better, and more useful, approach to generating the clusters.

The investigation of household electricity load profiles is an important area of research given the centrality of such patterns in directly addressing the needs of the electricity industry, both now and in the future. This work extends existing load profile work by taking electricity meter data streams and developing new ways of representing the household that can be used as the basis for clustering using existing data mining techniques. The identification of repeating motifs and the investigation of how the timing of the motifs varies from day to day, as a key behavioural trait of the household, is a novel area of research. An improvement in creating useful archetypes can have major financial and environmental benefits.

2   Methods and Technical Solutions

2.1   Load Profiling

There has been extensive research on determining daily load profiles to represent a household's electricity usage [6]. In many cases, (e.g., [7]), the daily load profiles are used as the basis for clustering "similar" households together to develop a small set of archetypal profiles which can be used for targeting of behaviour change interventions. Previous work has used different clustering techniques with the majority of the published literature using hierarchical clustering.

The common approach is to define a subset of the data (e.g. by season and/or by day of the week) and then to create average daily profiles for a household from the electricity meter data. The shapes of these daily profiles are then clustered to group similar shapes together. A representative profile is defined (e.g. by averaging all the members of the cluster) to produce a archetypal daily load profile for that cluster of households.

Previous work has not investigated how households may exhibit different behaviour from day to day and how these differences may be used as a distinguishing feature of the household and a basis for clustering.

2.2   Motifs

The electricity meter data reading stream from a household can be plotted as a graph of usage against time and regular activities appear as similar shaped patterns. Short patterns that repeat are defined as "motifs" and detection of these motifs, and their timing, can inform understanding of household behaviour.

This work uses the SAX (Symbolic Aggregate approXimation) technique which allows symbolic representation of time series data [8,9]. Other motif finding algorithms could also be incorporated into the proposed approach to identify the flexibility of behaviour (e.g. [10]). To assess variability within a household, it is necessary to detect the repeating motifs that are assumed to signify particular activities (e.g., cooking the evening meal). These are generally of a similar shape on different days but show some differences due to noise caused by other activities within the household (e.g., a fridge automatically running). The SAX approach of symbolising the real valued meter readings is useful as it allows for approximate matching (as various ranges of readings map to a single symbol).

Lines et al [11] applies motif finding to UK data to detect the use of particular appliances, drawn from a set of known appliances. This contrasts with the focus in this paper which is to find interesting, repeating patterns of behaviour without the need to define the activity that the motif represents. Appliances that can be consistently and accurately detected can be used with the approach detailed here by extending the analysis of the timing of repeating motifs to the analysis of variability of timing of appliance usage.

2.3   North East Scotland Electricity Monitoring Project (NESEMP)

This study makes use of data collected as part of the ongoing NESEMP which is examining the relationship between different types of energy feedback and psycho-social measures including individual environmental attitudes, household characteristics, and everyday behaviours. As part of this ongoing project, several hundred households are being monitored and the electricity usage is recorded every five minutes using CurrentCost monitors [12].

After removing data for households with insufficient readings, the data is loaded into a MySQL database and the readings are aligned with exact 5 minute boundaries (e.g. 1pm, 1.05pm, etc.) by interpolation between the actual readings. This is achieved by calculating the reading at an exact 5 minute point (e.g. 1.05pm) by considering the actual readings before and after that time and by calculating the reading such that the total usage over a longer period is the same whether the interpolated readings or the original actual readings are used [13]. This results in a set of 288 readings (one for every 5 minute period in the day) for each of the households in the database.

Each day of sampling is labelled in a number of ways such as "working day" or "summer" to aid selection of particular subsets of data.

2.4   Detecting Motifs

To find motifs within the data, each period of interest within the day (e.g., the peak period) for each household is examined by taking a moving window over the period. The subset of the meter readings within the moving window is then converted into a string and stored. Next, the window moves on by one time period (5 minutes) and the conversion into a string is repeated. Using an alphabet size of 5 and a motif size of 6 (i.e., 30 minutes), analysing the 4pm to 8pm period provides a total of 49 x 5 minute readings for each day. As the interest is in changes in usage rather than absolute usage, these readings are compared with adjacent readings in time to produce 48 values (one for each 5 minute period) representing the change in usage since the last 5 minute reading. This results in 42 motifs stored for each day for each household (one for each possible 30 minute period within the peak time). Fig. 1 shows an example of how the symbolised motifs are built up. The top graph shows the 5 minute readings for the 4 hour peak period. A sliding window of 6 readings (30 minutes) is taken across the peak period with the first 2 and the last window shown. Each window is normalised within the values in the window and then translated into the symbolised representations as shown at the bottom of the diagram.

The analysis uses an alphabet of 5 symbols (i.e., the letters "a" to "e") to represent the motifs. 5 is selected as a reasonable compromise between having too few symbols, and thus not detecting changes in electricity consumption, and having too many and thus generating too many patterns that do not repeat. The symbolisation translates readings within a particular range into a given letter and thus similar, although not identical, readings are translated into the same letter. The resulting motifs for 2 windows may be identical whereas the original readings may only be approximately similar.

The motif size selected is 6 corresponding to a 30 minute (i.e., 6 x 5 minutes) period. This figure was selected as the UK electricity settlement market uses a 30 minute period [14] and 30 minutes is also a reasonable period that will allow time for activities such as showering.

The motifs are built from the graph shape without regard to absolute value of the data. A possible effect of this is to find motifs within what is the general

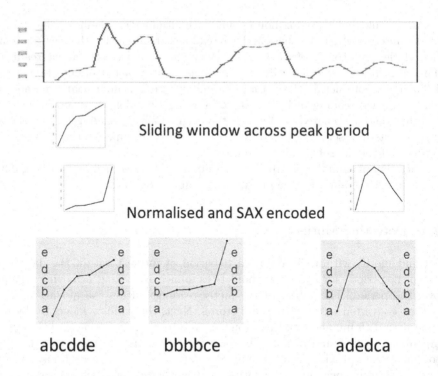

Fig. 1. Example of symbolisation (alphabet of 5, motif length of 6)

noise associated with the meter readings. This is avoided by ignoring any motifs within a window which have a range of less than 100W.

As the motifs are created by shifting a moving window over the stream of data, overlapping periods are considered and periods with no activity except for one change in meter reading will lead to a series of motifs that are similar. For example, a long period of no activity except for a jump of +200W will lead to motifs being found such as ccccca, ccccac, cccacc, etc. As only one of these is interesting for further analysis, the others are excluded.

The top motif (the one that occurs most often within a household) is further examined for the times when the motif occurs on each day. The number of times the motif occurs, and the standard deviation of the time of occurrence, are calculated for each household. Similarly, the second and third most common motifs within a household are identified and the variability in timing calculated.

Other useful measures relating to the motifs found within a household are also calculated including the number of different motifs (occurring at least twice) and the number of different motifs occurring on at least 30% of the days sampled for the household. The 30% figure is selected as a reasonable number to ensure only regularly repeating patterns are considered.

The attributes calculated for each household and used as input to the clustering algorithms are:

1. Number of occurrences of the motif occurring most frequently during the peak period.
2. Variability in timing of the occurrence of the most frequent motif within the household. This is represented by the standard deviation of the timing (measured in minutes) around the mean start time.
3. Number of occurrences of the second most frequent motif.
4. Variability in timing of the second most frequent motif.
5. Number of occurrences of the third most frequent motif.
6. Variability in timing of the third most frequent motif.
7. Total number of motifs for the household that occur at least twice.
8. Total number of different motifs that occur on at least 30% of days.

2.5 Clustering Algorithms

Various clustering techniques are selected for evaluation of the different approaches to analysing the data. Note that, whilst possibly a useful additional benefit, this work does not focus on selecting the "best" clustering algorithm but uses a selection of algorithms to assess the benefits or otherwise of making use of the motif variability information.

Based on the review by Chicco [6] the following clustering algorithms are selected as the most commonly used in previous work:

1. Kmeans is a well known algorithm that occurs in a number of examples of previous load profiling work. The algorithm requires a number of clusters (k) and works by randomly selecting an initial k locations for the centres of the clusters. Each data point is then assigned to one of the clusters by selecting the centre nearest to that data point. Once all the data points are assigned, each collection of points is considered, the new centre of the allocated points is calculated and the centre for that cluster is reassigned. The points are then reallocated to their new nearest centre and the algorithm continues until no changes are made to the allocations of points for an iteration [15].
2. Fuzzy c means. This provides an extension of the kmeans algorithm allowing partial membership to more than one cluster. The algorithm provides additional output showing the degree of membership that each household has of each of the derived clusters [16]. For this analysis, each household is assigned to the cluster for which they have the highest degree of membership.
3. Self Organising Maps. The Self Organising Map (SOM) is a neural network algorithm that can be used to map a high dimension set of data into a lower dimension representation. In this paper, the mapping is to a 2 dimensional set of representations which are arranged in a hexagonal map. Each sample (e.g., the average load profile for a given household) is assigned to a position in the map depending on the closeness of the sample to the existing nodes assigned to each position in the map (using a Euclidean measure of distance). Initially

the nodes are assigned at random but, over time, the map produces an arrangement where similar samples are placed closely together and dissimilar samples are placed far apart [17].

4. Hierarchical clustering. Most of the published load profiling work has used hierarchical clustering and this approach has the benefit of providing easily understood rules for cluster membership. The algorithm uses a dissimilarity matrix for the households and, starting initially with each household in its own cluster, proceeds by joining clusters which are most similar. The hierarchy is cut at a point to provide the desired number of clusters [18]. The Euclidean distance is used when creating the dissimilarity matrix and the Ward agglomeration method [19] is used for combining clusters. The Ward method minimises the sum of squares of possible clusters when selecting households to combine. Other agglomeration techniques tend to create a few small clusters containing extreme valued households plus one large cluster containing the remainder of the households.

5. Random Forests [20] is used to create a dissimilarity matrix which is used with Partitioning Around Medoids (pam) to form clusters. This is implemented using the R package randomForest [21].

A common issue is the appropriate setting for the number of clusters. To match common practice within the electricity industry, 8 clusters are selected. The UK electricity industry has worked with 8 load profiles since the 1990s [22]. Figueiredo et al [23] report that the Portuguese electricity utility aim for a number of clusters between 6 and 9.

2.6   Cluster Validity Measures

To assess the benefits of a particular cluster solution an appropriate cluster validity index needs to be used. Many have been considered in the literature with the Mean Index Adequacy (MIA) and the Cluster Dispersion Indicator (CDI) [24] used in most of the published load profiling work. Lower values for the CDI and MIA measures denote "better" solutions.

The data to be clustered consists of M records numbered as $m = 1, ..M$. Each record has H attributes numbered as $h = 1, ..H$. The hth attribute for the ith record is designated as $m_i(h)$.

The data is clustered into K clusters (numbered as $k = 1, .., K$). Each cluster has R_k members where $r_{(k)}$ is the rth record assigned to cluster k and $C_{(k)}$ is the calculated centre of the cluster k.

The distance (d) between 2 records is defined as:

$$d(m_i, m_j) = \sqrt{\frac{1}{H} \sum_{h=1}^{H}(m_i(h) - m_j(h))^2} \tag{1}$$

where $m_i(h)$ and $m_j(h)$ are the hth attributes for two records, m_i and m_j.

The "within set distance" $\hat{d}(S)$ of the members of a set, S with N members $(s_j$ where $j = 1, .., N)$ is defined as:

$$\hat{d}(S) = \sqrt{\frac{1}{2N} \sum_{n=1}^{N} \sum_{p=1}^{N} d^2(s_n, s_p)} \tag{2}$$

The MIA gives a value which relies on the amount by which each cluster is compact - i.e., if the members in the cluster are close together the MIA is low.

$$MIA = \sqrt{\frac{1}{K} \sum_{k=1}^{K} \sum_{r} d^2(r_{(k)}, C_{(k)})} \tag{3}$$

The CDI depends on the distance between the members of the same cluster (as for the MIA) but also incorporates information on the distances between the representative load diagrams (i.e., the centroids) for each cluster. This therefore measures both the compactness of the clusters and the amount by which each cluster differs from the others.

$$CDI = \frac{1}{\hat{d}(C)} \sqrt{\frac{1}{K} \sum_{k=1}^{K} \hat{d}^2(R_k)} \tag{4}$$

where C is the set of cluster centres and R_k is the kth cluster members set.

2.7 Processing

UK specific data is used to generate average daily load profiles for each household which are clustered to provide a baseline for comparison. Selected clustering algorithms are applied to the data and validity indexes are used to produce a measure of the quality of the partitions found.

Next, the novel approach of identifying motifs within the data, and measuring the variability in timing of the motifs, is used to generate a new set of derived data using the same UK dataset. The same clustering algorithms and validity indexes are then applied to this dataset. In addition, the results are compared with the baseline obtained from the average daily load profiles in the first step.

2.8 Assessing the Results

To assess the consistency of clustering solutions, the different arrangements of households into clusters are compared. The consistency of the clusters obtained from the different clustering algorithms is used as a measure of the quality of the results with more consistency between the results suggesting a more useful method of identifying the clusters.

Measuring consistency across the clustering results using the different sets of data (load profiles and motifs) may be criticised as not necessarily providing a true measure of quality as clustering results may be consistent but not necessarily represent useful, "true" clusters within the data.

The Rand index compares the different pairs of samples (i.e., each possible pair of households) and assesses the number in which each pair are in the same partition in the 2 different clustering solutions, the number where each member of the pair are in different partitions in both solutions, and the case where the members are in the same partition in one solution but a different partition in the other solution. The corrected Rand index [25] builds on the original work but adjusts the calculated value for the expected matching that would occur in a random arrangement. The corrected Rand index ranges from -1 to 1 with a higher value signifying better agreement between the partitions and hence a better solution.

3   Empirical Evaluation

3.1   Data Selection

A subset of the data is extracted for the peak period of 4pm to 8pm and for working days from Spring (March, April and May) 2011. Working days are weekdays excluding Scottish public holidays. Not all households have a full set of meter readings and those with less than 4 days of valid readings are excluded. The dataset has around 440,000 individual meter readings from 204 households.

The activities of interest within a household are related to switching appliances on or off (e.g., the use of electrical appliances in cooking) and it is the changes in the readings, rather than the absolute readings, that are of most interest and are used as the basis for analysis when using motifs.

3.2   Clustering Using the Load Profile Data

The data for the evening peak period (4pm to 7.55pm) are averaged to create a representative load profile for each household. For example, all the readings for 4pm for the household are averaged to create a representative reading for 4pm, similarly for 4:05pm, etc. The 204 representative profiles, each with 48 attributes (one for each time point), are then normalised within the 0 to 1 range and used with a variety of clustering algorithms.

3.3   Non-motif Variability Clustering

Various different measures of variability of behaviour within the household can be defined without the use of motifs (e.g., [26]) and two methods are considered.

One approach is to consider the time at which the maximum usage occurred on each day during the period of analysis. These times are then used to calculate the standard deviation of the time around the mean for each household.

A second approach is to consider the total usage during the peak period on each day during Spring 2011. The standard deviation of the total per day around the mean total per day also provides a measure of variability of behaviour. Each of the 2 measures are calculated and used as the basis of simple clustering using kmeans ($k = 8$). The households in each of the clusters are shown in Fig. 2.

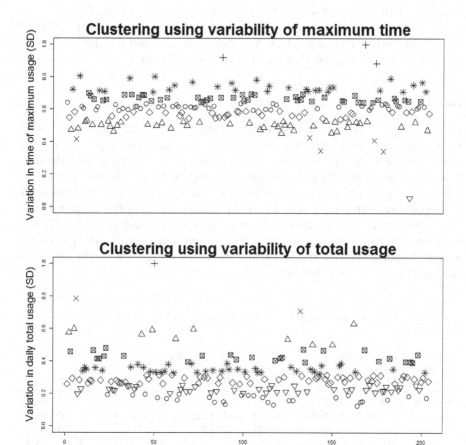

Fig. 2. Results for alternative variability measures

There is little correspondence between the cluster assignments for the 2 methods. The corrected Rand index of 0.01 shows no correspondence beyond that expected by chance. Furthermore, there is little correspondence with the clusters obtained from the motif variability approach (detailed below). A Spearman's rho value of 0.23 shows there is little correlation between the two different variability measures.

It is therefore concluded that neither of the non-motif measures give a useful, consistent measure of the variability of each household.

3.4   Clustering Using the Motif Data

This paper finds the motifs in the stream of meter data and then examines how the times of these repeating patterns vary from day to day within a household. Furthermore, the number of times a pattern repeats within a household is also used as an indication of the variability of behaviour of that household.

The motifs in the data are discovered and the attributes detailed in Section 2.4 are generated. The same clustering algorithms as used for the load profile clustering are then applied to produce 8 archetypal clusters.

3.5   Results

Various measures that represent the variability of behaviour can be constructed and this paper considers the variability in time of maximum usage and the variability in total usage. However, as each measure is intended to represent the same thing (i.e., the variability of behaviour), the fact that there is little correlation between the measures, or the membership of the clusters generated using the measures, means that they provide a poor representation of the characteristic.

Comparing the load profile results with the motif variability results, Table 1 shows, for each of the clustering algorithms used and for each set of data, the sizes of the partitions in the solution and the values for the MIA and CDI cluster validity indexes (lower is better).

Table 1. Clustering Results and Validity indexes

	Load Profiles			Motifs		
	Cluster sizes	MIA	CDI	Cluster sizes	MIA	CDI
Kmeans	10,16,19,20,20,27,44,48	0.593	1.34	2,5,7,26,29,37,41,57	0.445	0.641
Fuzzy	14,17,20,23,23,23,35,49	0.679	2.14	12,15,19,26,28,30,34,40	0.551	2.084
SOM	13,15,16,20,28,31,36,45	0.595	1.337	2,5,24,25,28,29,40,51	0.451	0.733
Hier	9,10,13,20,22,37,43,50	0.61	1.386	2,3,5,26,31,34,40,63	0.46	0.64
RF	14,18,19,28,29,29,30,37	0.794	1.131	18,18,19,21,25,32,35,36	0.628	1.34

The MIA and CDI values show that the kmeans and SOM techniques produce similar quality solutions using the load profiles. The hierarchical algorithm is less good with the Fuzzy Cmeans algorithm being significantly poorer. The random forest and pam combination provides a good result for the CDI measure but scores poorly on the compactness of the clusters (as measured by MIA).

When using the motif variability data, the kmeans, SOM and hierarchical algorithms produce similar quality results with the Fuzzy Cmeans algorithm again producing poorer results. The random forest and pam combination provides middling results.

Table 2. Modified Rand index of clusters using different clustering algorithms

	Profiles					Motifs				
	Kmeans	Fuzzy	SOM	Hier	RF	Kmeans	Fuzzy	SOM	Hier	RF
Kmeans	1	0.544	0.629	0.668	0.251	1	0.592	0.794	0.622	0.358
Fuzzy	0.544	1	0.562	0.491	0.355	0.592	1	0.626	0.511	0.447
SOM	0.629	0.562	1	0.49	0.287	0.794	0.626	1	0.591	0.33
Hier	0.668	0.491	0.49	1	0.272	0.622	0.511	0.591	1	0.312
RF	0.251	0.355	0.287	0.272	1	0.358	0.447	0.33	0.312	1

The MIA and CDI validity index calculations are not comparable between datasets due to the different number of attributes used.

Table 2 gives information on the consistency of the cluster partitions as the clustering algorithm changes. The results for the Rand index show that the values are consistently closer to 1 in the case of the clusters built using motif variation. The mean values for the Rand index (after omission of the values on the diagonal) are 0.4549 for the load profiles and 0.5183 for the motif variability approach. This shows a more consistent set of partitions are created when using the motif variability than the partitions created using the load profile information.

The results from the kmeans algorithm using the motif variability data can be seen at Fig. 3. The cluster with 26 houses shows very little variability in the timing of their regular activities and can be assumed to be "creature of habit" households who may not respond well to an incentive to change behaviour. The 2 house and the 29 house clusters show lots of repeating activities and may be best to target for interventions as there are likely to be many activities that often repeat and that may be modifiable.

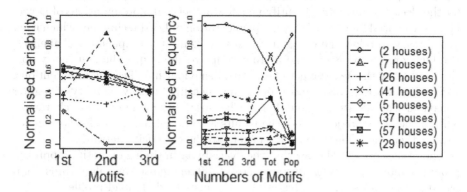

Fig. 3. kmeans clusters using motif variability

Examining the 29 house cluster in more detail, Fig. 4 show the motifs found for one of the houses and how the time of occurrence of the motif varies across

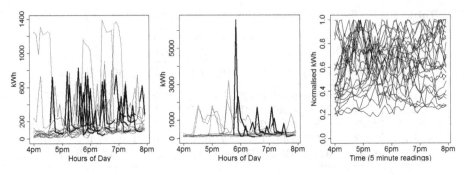

Fig. 4. Example house (high variability)

Fig. 5. Example house (low variability)

Fig. 6. Load profiles for high variability cluster

the 4pm to 8pm period. In contrast, the motifs for one of the houses in the 26 house cluster are shown in Fig. 5 and the timings can be seen to be less variable.

As a comparison, the average load profiles for each of the households in the 29 house cluster are shown at Fig. 6. There is little similarity between the households and hence, using the load profile shapes as the basis, little likelihood of the households being clustered together. However, the variability in timing of the motifs can be used as a method for selecting appropriate households to target and allows groupings to be designated as high or low variability.

4   Significance and Impact

The ability to cost effectively partition domestic households into a few meaningful archetypes based on the household electricity usage is an important problem for the electricity industry. Identifying a few archetypal representations of households is essential for cost effective implementation of DSM techniques which itself is necessary to allow the electricity industry to meet the upcoming challenges. Producing more consistent and more descriptive archetypes than currently possible will allow the deployment of effective behaviour modification interventions.

Previous work does not incorporate any measure of the variability of regular behaviour when clustering households. The variability is an important characteristic as one of the major uses of the results is to target incentives for households to vary their behaviour to provide benefit to the electricity network.

The results presented show that the "variability in timing of motifs" approach produces more consistent clusters across different clustering algorithms compared to the consistency of clustering using just the daily load profiles.

The symbolisation technique is effective in detecting repeating patterns (motifs) that are approximately the same shape. Depending on the type of intervention planned for a subset of the households (for example, incentives to change overall electricity usage from day to night, or to influence short periods of usage during the peak period), different sizes of motifs may be used.

This work shows a novel approach to using electricity meter data to cluster households that enhances and complements the existing techniques based on the daily load profiles.

Acknowledgements. This work was possible thanks to RCUK Energy Programme and EPSRC grant references EP/I000496/1 and EP/G065802/1 and forms part of the Desimax project [27].

Thanks are due to Pavel Senin for providing R code implementing the SAX method.

References

1. DECC: Towards a Smarter Future, Government Response to the Consultation on Electricity and Gas Smart Metering (2009)
2. River: Primer on demand-side management with an emphasis on price-responsive programs. Prepared for The World Bank by Charles River Associates, Tech. Rep. (2005)
3. Mooi, E., Sarstedt, M.: A concise guide to market research: The process, data, and methods using IBM SPSS statistics. Springer (2011)
4. Ramos, S., Figueiredo, V., Rodrigues, F., Pinhero, R., Vale, Z.: Knowledge extraction from medium voltage load diagrams to support the definition of electrical tariffs. International Journal of Engineering Intelligent Systems for Electrical Engineering and Communications 15(3), 143 (2007)
5. Ellegård, K., Palm, J.: Visualizing energy consumption activities as a tool for making everyday life more sustainable. Applied Energy 88(5), 1920–1926 (2011)
6. Chicco, G.: Overview and performance assessment of the clustering methods for electrical load pattern grouping. Energy 42(1), 68–80 (2012)
7. Ramos, S., Duarte, J., Soares, J., Vale, Z., Duarte, F.: Typical load profiles in the smart grid context - A clustering methods comparison. In: Power and Energy Society General Meeting, pp. 1–8. IEEE (2012)
8. Lin, J., Keogh, E., Wei, L., Lonardi, S.: Experiencing SAX: a novel symbolic representation of time series. Data Mining and Knowledge Discovery 15(2), 107–144 (2007)
9. Shieh, J., Keogh, E.: i SAX: indexing and mining terabyte sized time series. In: Proceeding of the 14th ACM SIGKDD International Conference on Knowledge Discovery and Data Mining, pp. 623–631. ACM (2008)
10. Mueen, A., Keogh, E., Zhu, Q., Cash, S., Westover, B.: Exact discovery of time series motifs. In: Proc. of 2009 SIAM International Conference on Data Mining, pp. 1–12 (2009)
11. Lines, J., Bagnall, A., Caiger-Smith, P., Anderson, S.: Classification of household devices by electricity usage profiles. In: Yin, H., Wang, W., Rayward-Smith, V. (eds.) IDEAL 2011. LNCS, vol. 6936, pp. 403–412. Springer, Heidelberg (2011)
12. Craig, T., Polhill, J.G., Dent, I., Galan-Diaz, C., Heslop, S.: The North East Scotland Energy Monitoring Project: Exploring relationships between household occupants and energy usage. Energy and Buildings (2014)
13. Dent, I., Craig, T., Aickelin, U., Rodden, T.: A Method for Cleaning and Storing Electricity Meter Data for Flexible Analysis. In: BeHave 2012, Helsinki (2012)

14. Elexon: The Electricity Trading Arrangements: A Beginner's Guide. Technical report, Elexon (2012)
15. Jain, A., Dubes, R.: Algorithms for clustering data. Prentice Hall College Div. (1988) Number 978-0130222787
16. Bezdek, J.C.: Pattern recognition with fuzzy objective function algorithms. Kluwer Academic Publishers (1981)
17. Kohonen, T.: The self-organizing map. Proceedings of the IEEE 78(9), 1464–1480 (2002)
18. Everitt, B.S., Landau, S., Leese, M., Stahl, D.: Cluster analysis. Edward Arnold, London (2001)
19. Ward Jr., J.H.: Hierarchical grouping to optimize an objective function. Journal of the American Statistical Association 58(301), 236–244 (1963)
20. Breiman, L.: Random forests. Machine Learning 45(1), 5–32 (2001)
21. Liaw, A., Wiener, M.: Classification and Regression by randomForest. R News 2(3), 18–22 (2002)
22. Electricity Association: Load profiles and their use in electricity settlement. UK-ERC (1997)
23. Figueiredo, V., Rodrigues, F., Vale, Z., Gouveia, J.: An electric energy consumer characterization framework based on data mining techniques. IEEE Transactions on Power Systems 20(2), 596–602 (2005)
24. Chicco, G., Napoli, R., Postolache, P., Scutariu, M., Toader, C.: Customer characterization options for improving the tariff offer. IEEE Transactions on Power Systems 18(1), 381–387 (2003)
25. Hubert, L., Arabie, P.: Comparing partitions. Journal of Classification 2(1), 193–218 (1985)
26. Dent, I., Craig, T., Aickelin, U., Rodden, T.: Finding the creatures of habit; Clustering households based on their flexibility in using electricity. In: Digital Futures, Aberdeen, UK (2012)
27. Kiprakis, A., Dent, I., Djokic, S., McLaughlin, S.: Multi-scale Dynamic Modeling to Maximize Demand Side Management. In: IEEE Power and Energy Society Innovative Smart Grid Technologies Europe 2011, Manchester, UK (2011)

Hybrid Recommender System for Prediction of the Yelp Users Preferences

Vladimir Nikulin

Department of Mathematical Methods in Economy,
Vyatka State University, Kirov, Russia
vnikulin.uq@gmail.com

Abstract. Recommender systems typically produce a list of recommendations in one of two ways - through collaborative or content-based filtering. Collaborative filtering approaches build a model from a user's past behavior (items previously purchased or selected and/or numerical ratings given to those items) as well as similar decisions made by other users; then use that model to predict items (or ratings for items) that the user may have an interest in. Content-based filtering approaches utilize a series of discrete characteristics of an item in order to recommend additional items with similar properties. These approaches are often combined, and called Hybrid Recommender Systems. In this paper we present hybrid recommender system, which was used online during ACM RecSys 2013 Contest, where we were awarded 2nd best prize. The contest was based on the real data, which were provided by Yelp - US internet based business recommender.

Keywords: collaborative filtering, content-based filtering, hybrid recommender systems.

1   Introduction

Recommender systems (RSs) attempt to profile user preferences over items, and model the relation between users and items. The task of recommender systems is to recommend items that fit a users tastes, in order to help the user in selecting/purchasing items from an overwhelming set of choices [1]. Such systems have great importance in applications such as e-commerce, subscription based services, information filtering, etc. Recommender systems providing personalized suggestions greatly increase the likelihood of a customer making a purchase compared to unpersonalized ones. Personalized recommendations are especially important in markets, where the variety of choices is large, the taste of the customer is important, and last but not least the price of the items is modest. Typical areas of such services are mostly related to art (esp. books, lectures, movies, music), fashion, food and restaurants, gaming and humor.

Recommender systems play an important role in such highly rated Internet sites as Amazon.com, YouTube, Netflix, Yahoo, Tripadvisor, Last.fm, and IMDb. Moreover many media companies are now developing and deploying RSs as part

P. Perner (Ed.): ICDM 2014, LNAI 8557, pp. 85–99, 2014.

of the services they provide to their subscribers. There are dedicated conferences and workshops related to the field. We refer specifically to ACM Recommender Systems (RecSys), established in 2007 and now the premier annual event in recommender technology research and applications [2]. At institutions of higher education around the world, undergraduate and graduate courses are now dedicated entirely to recommender systems; tutorials on RSs are very popular at computer science conferences; and recently a book introducing RSs techniques was published [3]. There have been several special issues in academic journals covering research and developments in the RS field. Among the journals that have dedicated issues to RS are: AI Communications (2008); IEEE Intelligent Systems (2007); International Journal of Electronic Commerce (2006); International Journal of Computer Science and Applications (2006); ACM Transactions on Computer-Human Interaction (2005); and ACM Transactions on Information Systems (2004).

Content-Based systems focus on properties of items. Similarity of items is determined by measuring the similarity in their properties.

Collaborative-Filtering systems focus on the relationship between users and items. Similarity of items is determined by the similarity of the ratings of those items by the users who have rated both items.

Suppose that user is new to the market ("cold start"), and we have no any corresponding review records. In this case, and in accordance with the content-based approach, we can use available properties for user and business in order to predict how user will rate the business under consideration.

In recommender systems, the cold start problem is often reduced by adopting a hybrid approach between content-based matching and collaborative filtering [4]. New items (which have not yet received any ratings from the community) would be assigned a rating automatically, based on the ratings assigned by the community to other similar items. Item similarity would be determined according to the items' content-based characteristics such as empirical probabilities or association rules. Note that traditional data mining techniques such as association rules were tried with good results at the early stages of the development of recommender systems [5].

Recent research has demonstrated that a hybrid approach, combining collaborative filtering and content-based filtering could be more effective in some cases. Hybrid approaches can be implemented in several ways: by making content-based and collaborative-based predictions separately and then combining them; by adding content-based capabilities to a collaborative-based approach (and vice versa); or by unifying the approaches into one model. Several studies empirically compare the performance of the hybrid with the pure collaborative and content-based methods demonstrate that the hybrid methods can provide more accurate recommendations than pure approaches. These methods can also be used to overcome some of the common problems in recommender systems such as cold start and the sparsity problem.

It is not enough to create vectors describing items or businesses, we, also, have to create vectors with the same components that describe the users preferences.

We have the list of connections between users and items (it maybe list of review rates or list of purchases, for example). With this information, the best estimate, we can make regarding which items the user likes, is some aggregation of the profiles or properties of those items.

Note that the number of sufficiently frequent properties or categories maybe very large (in our case, see Section 2.2, it is 354). Accordingly, it will be too difficult to load and analyse the data in the standard form of the tables. One way to overcome this problem would be transfer all data (including locations of the businesses) into the sparse format. However, we decided to implement special novel transformation, which in described in Section 5. This transformation let us keep all the remaining blocks in a standard form. In order to calculate the final predictions, we used homogeneous ensembling [6], where any single GBM-based learner in R was based on about 10% of all data. In line with the main computations, we were able to validate our model with CV-passports, see Remark 8.

2   ACM RecSys 2013 Yelp Data

The Contest was started on 24th April 2013 and was ended on 31st August 2013 (129 total days). Initial period of the contest was ended on 24th August with 465 actively participated teams. The final solutions for the problem were submitted by 158 teams, and our solution was formally recognised as the 2nd best, see Table 4, where top ten teams and results are presented.

RecSys database[1] was given in JSON format, and includes two parts 1) training with 229907 known reviews (integers from 1 to 5 - assessments of the businesses by customers or users), where 1 stands for poor and 5 - for excellent; 2) testing with 36404 unknown reviews (to be predicted). The task was to minimise RMSE - the root mean squared error:

$$RMSE = \sqrt{\frac{\sum_{i=1}^{n}(\hat{s}_i - s_i)}{n}}, \tag{1}$$

where n is the total number of review ratings to be predicted, $\hat{s}_i$ is predicted value for review i, s_i is actual rating for review i.

Both datasets are divided into four groups (see Table 1): 1) business (or bus - to simplify notations), 2) checkins (statistical information about day of the week and time in hours when enquiry regarding related business was submitted), 3) users and 4) reviews.

All four groups are divided into subgroups: for training and testing. Note that train.bus ∩ test.bus = ∅; train.user ∩ test.user = ∅, where train.reviews are formulated in the terms of train.bus and train.users, test.reviews are formulated in the terms of all data: train.bus, test.bus, train.user and test.user. The coverage of the test.reviews by the train.bus and test.bus is full and complete, but in the significant proportion of all cases (2039 out of 36404) users are absolutely new.

---

[1] http://www.kaggle.com

Table 1. Some statistical characteristics of the RecSys 2013 Challenge data

name	train	test
business	11537	2797
checkin	8282	1796
user	43873	9522
review	229907	36404

That means, they can not be found in any of the available two sets: train.users or test.users. Such problem is known as "cold start" [7]-[8].

2.1   Additional Information

The data, which were described in the previous section, are sufficient in order to implement collaborative filtering. In this section we shall describe some additional information, which was provided by the organisers, and maybe used as a base for the content-based modelling.

The following information was given for train.bus: 1) city, 2) state, 3) latitude, 4) longitude, 5) $s(b)$ - average stars (star ratings or average reviews, rounded to half-stars), 6) $r(u)$ - review counts, 7) categories (see Table 2); and for train.users: 1) $r(u)$ - review counts, 2) $s(u)$ - average stars, 3) votes ('useful', 'funny', 'cool').

Remark 1. Note that stars field was removed from the test.bus; stars and votes fields were removed from test.users.

2.2   Business Categories

As an illustration, let us consider 13 popular businesses, as they maybe seen on the Yelp web-site: Arizona, Bank of America, Credit Union West, Kmart, Pizza-Hut, Jimmy John's, Starbucks, Panda Express, Homewood Suites (Hilton), Target, Safeway, McDonald's and Walmarts.

We counted $n_c = 354$ sufficiently frequent categories, which were used in train.business not less than 5 times. Other categories were ignored.

Remark 2. The categories data are sparse. This property is a very essential: not more than ten categories are used for the description of any business.

Further, we formed binary matrix of features $\mathcal{A}$ of $\{n_b \times n_c\}$, where $n_b = 11537$ - the size of train.bus set.

Stars (or average business ratings) were used as a target variables and categories were used as features (binary data, where one stands for present, and zero - for absent). Using "randomForest" function in R, we computed positive importance RF-ratings of different categories (bigger value means greater importance). The size of the training data in terms of reviews is a quite large. Accordingly, it will be very important to reduce dimensionality of the data, and we shall consider this topic in Section 5.

Table 2. Examples of categories for some businesses, where zero means empty space

Food	Ice Cream + Frozen Yogurt	0
Pet Services	Pet Boarding/Pet Sitting	Pets
Active Life	Yoga	Fitness + Instruction
Hobby Shops	Shopping	Toy Stores
Bars	Nightlife	0
Department Stores	Fashion	Shopping
Pizza	Restaurants	0
Mexican	Restaurants	0
Tanning	Beauty + Spas	0
Auto Repair	Automotive	0
Professional Services	0	0
Greek	Restaurants	0
Food	Donuts	Coffee + Tea
Hotels & Travel	Event Planning + Services	Hotels
Chinese	Restaurants	0
Local Services	Appliances + Repair	Home Services
Auto Repair	Automotive	0
Food	Coffee & Tea	0
Banks & Credit Unions	Financial Services	Mortgage Brokers
Department Stores	Fashion	Shopping
Hotels & Travel	Event Planning + Services	Hotels
American (New)	Restaurants	0
Breakfast & Brunch	Restaurants	0
Tapas/Small Plates	Restaurants	0
Health & Medical	Dentists	General Dentistry
Arts & Entertainment	American (Traditional)	Music Venues

3   Model N1: Content-Based Filtering with Average Stars

This model is the most straightforward, but complete component of the proposed recommender system.

Suppose, we would like to calculate prediction $\hat{s}(u, b)$ how user u will rate business b, where the pairs $\{s(u), r(u)\}$ and $\{s(b), r(b)\}$ are known (that means, both u and b are listed in the train.user and train.business). Then, we can implement the following formula

$$\hat{s}(u, b) = \exp(\frac{r(u) \cdot \log(s(u)) + r(b) \cdot \log(s(b))}{r(u) + r(b)}). \qquad (2)$$

Otherwise, in the cases if $s(u)$ or $s(b)$ are known, we shall apply the following formulas

$$\hat{s}(u, b) = s(u); \qquad (3a)$$
$$\hat{s}(u, b) = s(b). \qquad (3b)$$

Table 3. List of the most important categories, where n is the number of occurrences in the train.bus set

Index	Category	n	Rating
73	Restaurants	4503	46.6809
211	Fast Food	386	34.6272
216	Apartments	83	30.4125
138	Real Estate	118	25.6183
28	Mobile Phones	70	24.2056
212	Hotels + Travel	379	17.0649
324	Specialty Food	181	15.8686
24	Beauty + Spas	764	14.3454
301	Hotels	284	13.0453
293	Active Life	525	11.0247
175	Food	1616	10.6413
93	Internet Service Providers	22	10.6279
81	Hair Salons	154	10.6242
160	Nail Salons	242	10.5391
247	Home Services	409	10.0099
232	American (Traditional)	480	9.7582
88	Shopping	1681	9.5534
268	Drugstores	125	9.1458
126	Event Planning + Services	453	9.0172
264	Chiropractors	39	8.5157
11	Car Wash	70	8.4186
29	Performing Arts	54	7.8904
244	Health + Medical	471	7.7370
343	Fitness + Instruction	204	7.3313
166	Professional Services	71	7.2573

3.1 Model N2 (Case of Star Averages, Extracted from the Review Data)

Using train.review data we can compute average stars for all involved businesses and users. In addition, we compute the numbers of reviews (support): $\phi(u)$ and $\phi(b)$. Further, we distinguish users and businesses according to the numbers of reviews.

Smoothing with Average Stars as Functions of the Numbers of Reviews. Clearly, average stars, which are based on small numbers of reviews are noisy and should be regularised. On the other hand, the numbers of users and businesses with small number of reviews are large (see Figure 1(a,c)), and the corresponding average stars z (see Figure 1(b,d)) maybe applied as regularises.

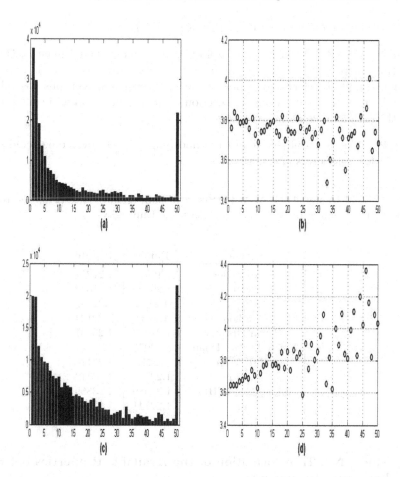

Fig. 1. (a) support for users, see Section 3.1; (b) average stars for users; (c) support for businesses; (d) average stars for businesses, where all horizontal axes represent number of occurrences in the training set

Remark 3. It is very interesting to note that users with bigger number of reviews appears to be more conservative (there is a slight decline in value of average stars $z(\phi(u))$, see Figure 1(b)). On the other hand, we can see that businesses which are more popular and attract greater attention (means with bigger number of reviews) are gaining higher average stars $z(\phi(b))$, see Figure 1(d) (smoothing parameters 3 and 6 were used for users and businesses to compute average stars).

The following smoothed averages were used as an input in (2)

$$\tilde{s}(u) = (1 - \alpha)\hat{s}(u) + \alpha z(\phi(u)),\qquad (4)$$

with smoothing parameter

$$\alpha = \frac{1}{(1 + \phi(u))^{\gamma}},$$

where $\gamma = 0.25$ (computation of $\tilde{s}(b)$ is an identical).

Remark 4. The ACM RecSys 2013 Yelp lasted for more than 4 months. During that time there were two main opportunity to obtain a feedback: 1) external, through Kaggle LeaderBoard, and 2) internal, using homogeneous ensembling with CV-passports, see Section 4. Selection of the parameters was a consequence of both feedbacks, combined together.

Remark 5. The main objective of the smoothing (4) is to overcome "cold star" problem for both users and businesses.

Table 4. Final results of the ACM RecSys 2013 data mining contest: top ten teams, where "vsu" is our team (stands for Vyatka State University)

N	Team	Public	Private
1	BrickMover	1.19826	1.21251
2	**vsu**	1.20456	1.21552
3	Sergey	1.2112	1.22856
4	Merlion	1.2154	1.22724
5	Bryan Gregory	1.21923	1.23107
6	Gxav & Paul Duan	1.22078	1.23021
7	Biro Biro & Dmitry	1.2212	1.23118
8	YaTa	1.22518	1.23714
9	Xiong Cao & HIT	1.22589	1.23869
10	Li	1.22741	1.23699

3.2 Model N3: Transformation of the Available Properties between Users and Businesses

Business categories, which are available for both train and test parts of the data, represent key elements in the proposed recommender system. Using relations between users and businesses (which are represented by the reviews), where star ratings are not necessary, we transfer categories to users (compute sums of categories). Plus, we compute for users averages in the terms of latitudes and longitudes (locations of businesses). As a next step, we found expression of any category in the terms of votes and checkins (week-day and time). Further, we calculated expression of any user and business in the terms of votes and checkins, see Sections 5.1 and 5.2. It was noticed that $\hat{s}(user, business)$ - user's rating (named, also, as a star) of the business, maybe predicted with high quality in the case if we know average stars and numbers of reviews for both input parameters: user and business. In this particular project, we considered prediction of average stars (numbers of reviews are available) for users and businesses, using as an explanatory variables or features 1) locations (latitudes and longitudes), 2) numbers of reviews, 3) categories, 4) votes (processed by two different methods) and 5) checkins. Finally, we input all above data for users and businesses into train.review and test.review sets, where we used predictions of the average stars for the test.users and test.businesses. As a predictor we applied GBM [9] and randomForest [10] functions in R.

Remark 6. The model, described in this section, represents a form of unsupervised learning. There are many cases when user from the test.review is not given in train.user or in test.user sets. However, we can obtain properties (in accordance with test.review), and use those properties as an input to the GBM predictor.

3.3   Computation of the Final Solution: Hybrid Model as an Ensemble of the Models NN 1, 2 and 3

We can ensemble the outcomes of the models N1 and N2 according to the formula (2), where the numbers of reviews equal to $r(u) + r(b)$. Note that $r(u)$ and $r(b)$ are different for the models N1 and N2. As a consequence, we shall produce ens_{12}.

Let us denote by $q_i = r_i(u) + r_i(b), i = 1, 2$, where $i = 1$ corresponds to the model N1, and $i = 2$ corresponds to the model N2.

Then, the final solution is calculated according to the formula

$$ens_{\text{final}} = \theta ens_{12} + (1 - \theta)ens_{\text{gbm}}, \tag{5}$$

where

$$\theta = \left(\frac{10}{q_1 + q_2 + 1} \right)^{\lambda}, \lambda = 0.2,$$

subject to the condition that $\theta \leq 1$, ens_{gbm} - heterogeneous ensemble as an outcome of the model N3.

Remark 7. In order to produce ens_{gbm} we used about 30 of different homogeneous ensembles, where each of which was based on the different configuration of the database. In particular, we conducted experiments with date (considering test.reviews as a future), gender (according to the name) and city.

Remark 8. In line with computation of the homogeneous ensembles as it was described in [6] (see, also, Section 4), we were able compute the corresponding CV-passports. The following results we observed in the terms of CV-passports and training error: 1) review.stars predictions $\{0.957, 0.8738\}$, 2) user.stars predictions $\{0.94, 0.852\}$, and 3) business.stars predictions $\{0.7838, 0.7049\}$. As an output, our model produces not only test.predictions, but, also, train.predictions, which maybe used for smoothing (similar to (4)).

In fact, we considered, also, model N4 - matrix factorisation via stochastic gradient descent [11], but it did not produce any significant improvement.

Table 5. Categorical data in a novel format (see Section 5), where any row corresponds to the particular business, "v11" - number of different categories. Columns from 1 to 10 are sequenced in a decreasing order according to the importance.

v1	v2	v3	v4	v5	v6	v7	v8	v9	v10	v11
175	305	305	305	305	305	305	305	305	305	2
278	299	43	43	43	43	43	43	43	43	3
293	343	312	312	312	312	312	312	312	312	3
0	0	0	0	0	0	0	0	0	0	0
167	53	53	53	53	53	53	53	53	53	2
88	213	352	352	352	352	352	352	352	352	3
73	341	341	341	341	341	341	341	341	341	2
73	182	182	182	182	182	182	182	182	182	2
24	111	111	111	111	111	111	111	111	111	2
242	303	303	303	303	303	303	303	303	303	2
166	166	166	166	166	166	166	166	166	166	1
73	260	260	260	260	260	260	260	260	260	2
175	329	47	47	47	47	47	47	47	47	3
126	212	301	301	301	301	301	301	301	301	3
73	98	98	98	98	98	98	98	98	98	2
247	152	192	258	258	258	258	258	258	258	4
242	303	303	303	303	303	303	303	303	303	2
175	329	329	329	329	329	329	329	329	329	2
247	138	65	308	30	30	30	30	30	30	5
88	213	352	352	352	352	352	352	352	352	3
126	212	301	301	301	301	301	301	301	301	3
73	85	85	85	85	85	85	85	85	85	2
73	156	156	156	156	156	156	156	156	156	2
73	73	73	73	73	73	73	73	73	73	1
244	210	70	58	58	58	58	58	58	58	4
73	167	53	232	202	236	236	236	236	236	6
88	213	137	224	224	224	224	224	224	224	4
247	138	216	216	216	216	216	216	216	216	3
126	212	301	301	301	301	301	301	301	301	3
73	286	107	1	1	1	1	1	1	1	4
247	138	30	30	30	30	30	30	30	30	3

Using an External Data from the Yelp Web-Site. We note that some business star averages maybe easily collected manually from the Yelp web-site[2]. It is a fairly standard and publicly available procedure, which require no any special knowledge or skills. After that those averages maybe used in the model N1 of the proposed method. That means, the corresponding businesses will be transferred from test.business to train.business. Consequently, we shall observe some improvement (about 0.004) of the score in the terms of the root mean squared error. We do believe that some limited data warehousing (work with and collection of the real data) as an essential element of the Contest is highly

---

[2] http://www.yelp.com/phoenix

Table 6. Original checkin data (15 features), see Section 5.2

1	2	3	4	5	6	7	8	9	10	11	12	13	14	15
0	5	22	24	7	26	0	16	33	8	0	30	0	0	20
0	9	45	36	0	58	0	30	64	0	0	38	0	0	2
2	21	32	42	3	48	0	51	40	6	0	34	0	0	5
0	3	41	23	1	44	0	30	80	2	0	48	0	0	12
4	2	20	14	6	10	0	18	24	9	0	42	0	0	40
0	6	49	30	4	37	0	24	72	4	0	44	0	0	13
0	2	22	19	4	21	0	20	36	5	0	43	0	0	17
0	3	24	18	4	24	0	19	39	4	0	38	0	0	16
0	4	23	17	2	34	0	22	47	2	0	32	0	0	6
0	8	28	22	1	46	0	12	50	1	0	18	0	0	2
0	1	15	7	0	18	0	8	26	0	0	19	0	0	3
0	2	27	15	2	32	0	18	48	2	0	33	0	0	11
1	27	38	42	2	76	0	14	50	3	0	23	0	0	9
4	17	34	35	9	42	0	20	46	9	0	43	0	0	26
0	2	25	20	4	24	0	20	38	3	0	39	0	0	13

desirable in some cases, as it may help participants to understand the data better. For example, the winner of the PAKDD 2010 data mining Contest[3] used external data, and it was highly rewarded. Generally, it is a good idea to encourage some initiative. In fact, it is far from easy to find out what sort of data to collect, where to find those data, and how to use new data in the model.

4   Calculation of the CV-Passports as a Validation Trajectories against All Training Data

In accordance with the principles of homogeneous ensembling, it appears to be natural to consider calculation of the decision function as an average of the large number of the single learners (or base classifiers), where any single learner is based on the randomly selected subsamples of observations and features.

Let us describe the proposed method in more details. Suppose that we are using about 75% of the available data for training. Then, remaining 25% of the data maybe used for the validation control. In line with the principles of cross-validation, we shall test stability of the validation results by considering a sequence of the random splittings.

Definition 1. *We are proposing to accumulate the validation results against all training samples in line with construction of the homogeneous ensemble. We shall call averaged validation trajectory as a CV-passport of the corresponding homogeneous ensemble.*

---

[3] http://sede.neurotech.com.br/PAKDD2010/

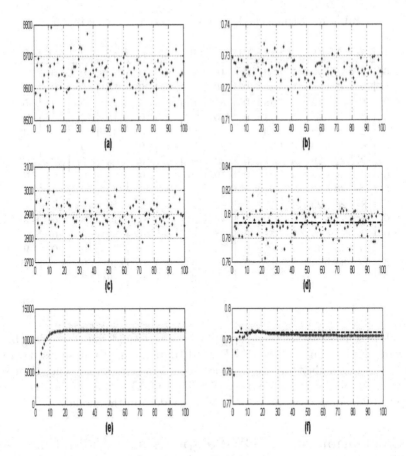

Fig. 2. Homogeneous ensembling applied to estimation of the star ratings of the test businesses, where horizontal axis determines index of random splitting. Left column shows numbers of samples used for (a) training, (c) testing and (e) calculation of CV-passport. Right column shows the corresponding errors in terms of RMSE (1).

Figure 2(d) illustrates the process of standard cross validation, where dashed black line corresponds to the average after 100 random splittings. It is interesting to note that all training samples will be filled after about 25 splittings, see Figure 2(e). We note, also, that evaluation results with CV-passports are slightly better compared to the the standard CV, see Figure 2(f).

5   Dimensional Reduction (Novel Format for Categorical Variables)

This section represents one particular component of the proposed method as a main novelty of the proposed recommendation system. We replaced ones in the

Table 7. Dimensional reduction: data of Table 6 after transformation (only 5 features)

9	12	6	4	3
9	6	3	12	4
8	6	4	9	12
9	12	6	3	8
12	15	9	3	8
9	3	12	6	4
12	9	3	6	8
9	12	6	3	8
9	6	12	3	8
9	6	3	4	12
9	12	6	3	8
9	12	6	3	8
6	9	4	3	2
9	12	6	4	3
12	9	3	6	8

matrix $\mathcal{A}$ by the corresponding positive RF-rating, zeros were kept intact. After that, we sorted any row in a decreasing order, and replaced positive values by the corresponding indexes (see Table 5). By definition, there are 11 columns in the secondary matrix of indexes $\mathcal{B}$, where first column contain indexes of the most important categories, second column contain indexes of the less important categories if available and so on. The maximum number of the different indexes is 10 (see Remark 2), and this number is given in 11th column. In the case if we don't know any categories for this business, all values in the corresponding row of the matrix $\mathcal{B}$ are set to zero.

5.1 Transfer of the Votes Data to all Users and Businesses

We have "votes = (useful, funny, cool)" information for train.users, and the task is to transfer this knowledge to the remaining data. As it was described in Section 3.2, we can find expression of all users in the terms of categories. Then, we can compute matrix $\mathcal{C}$ of [categories, votes], using data from train.users. After that, we can explain all users and businesses in the terms of votes, and compute related secondary features: for example, normalised vectors of votes or sums of votes, log (sums) and so on.

5.2 Transfer of the Checkin Data to All Users and Businesses

The idea to transform checkin data (hour, day of the week, number of counts) to all users and businesses is about the same as in the case of votes, see above Section 5.1. Note that checkin information is not available for all businesses in the both train and test sets.

Firstly, we applied some smoothing. We transferred 7 days of the week to only 3 different categorical values: $\{1, 2, 2, 3, 3, 3, 1\}$. Also, we split 24 hours into

5 sub-intervals: $\{0 : 5, 6 : 9, 10 : 15, 16 : 19, 20 : 23\}$. The product of 3 and 5 will give us 15 new features. Then, we computed matrix $\mathcal{D}$ of $[n_c, 15]$, $n_c = 354$ categories and 15 checkins, using data from train and test businesses taking into account the numbers of checkins. After that, we can explain all users and businesses in the terms of checkins, see Table 6. Further, we applied dimensional reduction from 15 to 5 features according to the method similar to Section 5. Figure 2 illustrates the process of homogeneous ensembling for the estimation of the star ratings of the test businesses, applied to the database containing two main blocks: with 11 features as described in Section 5 and with 5 features as described in this section. Besides, we included 1) count numbers (for star ratings) and 2) geographical location with latitude and longitude.

6   Concluding Remarks

The proposed system represents an ensemble of three models, where the first one is content-based, the second one maybe classified as collaborative filtering, and the last one is a hybrid. We have found that the third model is the most influential, and produced very significant improvement. This model is based on the following two facts: 1) all businesses in review data maybe found in train.bus or test.bus; 2) all businesses are categorised. Further, we can transfer categories (plus some additional information) to all the users in review data. Consequently, we shall produce standard regression model with rectangular matrix of explanatory variable. Finally, it is proposed to solve a high-dimensionality problem using novel method as described in Section 5. Note that this method is general, and maybe applied for the variety of similar tasks.

References

[1] Takacs, G., Pilaszy, I., Nemeth, B., Tikk, D.: Scalable collaborative filtering approaches for large recommender systems. Journal of Machine Learning Research 10, 623–656 (2009)
[2] Ricci, F., Rokach, L., Shapira, B., Kantor, P. (eds.): Recommender systems handbook. Springer (2011)
[3] Jannach, D., Zanker, M., Felfernig, A., Friedrich, G. (eds.): Recommender systems an introduction. Cambridge University Press (2010)
[4] Schein, A., Popescul, A., Ungar, L., Pennock, D.: Methods and metrics for cold-start recommendations. In: Proceedings of the 25th Annual International ACM SIGIR Conference on Research and Development in Information Retrieval (SIGIR 2002), pp. 253–260. ACM, New York City (2002)
[5] Agrawal, R., Srikant, R.: Fast algorithms for mining association rules. In: Proc. 20th Int. Conf. Very Large Data Bases, 32 p. (1994)
[6] Nikulin, V., Bakharia, A., Huang, T.-H.: On the evaluation of the homogeneous ensembles with CV-passports. In: Li, J., Cao, L., Wang, C., Tan, K.C., Liu, B., Pei, J., Tseng, V.S. (eds.) PAKDD 2013 Workshops. LNCS, vol. 7867, pp. 109–120. Springer, Heidelberg (2013)

[7] Spyromitros-Xioufis, E., Stachtiari, E., Tsoumakas, G., Vlahavas, I.: A hybrid approach for cold-start recommendations of videolectures. In: Smuc, T., Antulov-Fantulin, N., Morzy, M. (eds.) ECML/PKDD Workshop and Conference Proceedings, Discovery Challenge, pp. 29–40 (2011)

[8] Iaquinta, L., Semeraro, G.: Lightweight approach to the cold start problem in the video lecture recommendation. In: Smuc, T., Antulov-Fantulin, N., Morzy, M. (eds.) ECML/PKDD Workshop and Conference Proceedings, Discovery Challenge, pp. 83–94 (2011)

[9] Friedman, J., Hastie, T., Tibshirani, R.: Additive logistic regression: a statistical view of boosting. Annals of Statistics 28, 337–374 (2000)

[10] Breiman, L.: Random forests. Machine Learning 45, 5–32 (2001)

[11] Nikulin, V., Huang, T.-H.: Unsupervised dimensionality reduction via gradient-based matrix factorization with two learning rates and their automatic updates. Journal of Machine Learning Research, Workshop and Conference Proceedings 27, 181–195 (2012)

Analysis Using Popularity Awareness Index, Recency Index and Purchase Diversity in Group Buying

Yasuyuki Shirai[1], Hiroyuki Morita[2], Masakazu Nakamoto[1], and Satoshi Oyama[3]

[1] JST-ERATO Minato Discrete Structure Manipulation System Project,
Hokkaido University, Sapporo, Japan
{shirai,nakamoto}@erato.ist.hokudai.ac.jp
[2] College of Sustainable System Sciences,
Osaka Prefecture University, Osaka, Japan
morita@eco.osakafu-u.ac.jp
[3] Graduate School of Information Science and Technology,
Hokkaido University, Sapporo, Japan
oyama@ist.hokudai.ac.jp

Abstract. We propose new metrics for customers' purchasing behaviors in a group buying coupon website, based on HITS algorithms and information entropy: that is, popularity awareness index, recency index, and purchase diversity. These indices are used to classify customers and predict future behaviors. This paper includes definitions of these new indices to be used in real group buying websites. In these websites, adequate characteristics for customers are strongly required and are critical for marketing purpose. We will also provide some experimental results on real data set, including customer segmentation used in future marketing planning.

Keywords: group buying, HITS algorithm, entropy.

1 Introduction

Various types of e-commerce have grown steadily around the world. In Japan, the domestic business-to-consumer (B-to-C) e-commerce market size has reached 9.5 trillion yen in 2012 [1]. In addition, the business models of e-commerce transactions have diversified accordingly, with the emergence of group buying coupon sites. In this model, retailers exhibit a coupon in the group buying coupon site restricting purchase time limit, conditions of purchase, and so on. In addition, the retailers set a limit to the minimum quota of transactions. For instance, if total volume of coupons purchased by customers does not exceed the minimum volume, the deal will not be established. Although setting the minimum volume does not seem to facilitate sales, customers who want to purchase the coupons would refer their friends to sign up for the coupon in order to propagate enough quota to establish the transaction within the purchase time limit. This

P. Perner (Ed.): ICDM 2014, LNAI 8557, pp. 100–114, 2014.

framework is referred as flash marketing. GROUPON and PONPARE are two big group buying coupon websites in Japan, with coupon sales averaging at 1.5 billion yen every month [2]. However, the market share changes rapidly, thus every site has a chance to lead the market. In order to outperform against other rival companies, identification of proper customer segmentation and adequate implementation of Customer Relationship Management strategy are necessary.

In this paper, we analyze the web log data and related data provided from a group buying coupon site in Japan[1]. The data consists of one year of transactions from July 2011 to June 2012. Although the behavior and purchased history which can be used for analysis are sampled coarsely to preserve privacy, personal attributes including age and gender are distributed. Information about the coupon, namely, the name of the coupon, fixed price, purchased price, discount rate, genre of the coupon are also available. From our preliminary analysis of the data, we have noted the following attention points.

1. The stakeholders of this site are the site managing company, retailers, and purchasers. And the retailers exhibit only one coupon at one time without repeating the exhibition.
2. The data is coarsely-sampled as we mentioned. Even if we aggregate total number of coupons purchased, we will not be able to grasp the entire behavior of the customers.
3. Most of the coupon discount rates are more than 50% (high level of discount). Moreover, there are several coupons whose discount rates are 100%.

From the above observations, it is difficult to focus on both the price and the continuous selling in our analysis. Instead we think that it is more important to analyze the relationship between customers and the coupons. Purchasing behaviors can be classified into various types. Some purchasers may only be interested in popular coupons among others purchasers, and some purchasers may be interested in unique coupons.

We think that focusing on the popularity of the coupon with such purchasing characteristics will be very insightful. Thus, we have defined coupon popularity awareness index for analyzing the customer's behavior based on the index. To do this, we propose using Hypertext Induced Topic Selection algorithm (HITS) [5] to define the index. HITS calculates hub store and authority score between websites from web log data and derives the importance of the website of the search engine. Although the relationship among websites can be expressed by general graph, the relationship between customers and coupons can be expressed by bipartite graph. We apply HITS algorithm to such bipartite graph and calculate the hub and authority score in the same way as usual HITS. In this paper, we shall define the scores as coupon popularity index and popularity awareness index, respectively.

Additionally, we focus on the customers' response when making purchases. We further defined the customers' response to the release of coupons on the

[1] The data is provided by Joint Association Study Group of Management Science at Data Analysis Competition in 2012.

website : for example, some customers purchase coupons immediately after the release, while some customers purchase coupons after observing other customers' behaviors. Figure 1 illustrates the relationship between the elapsed time after the release and the sales volume[2]. We discovered that the highest volume of

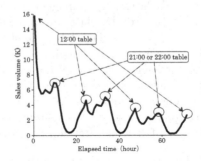

Fig. 1. Transition of purchasing from the coupon release

purchase takes place during the first two hours from the release. The cycle is repeated in 24 hour intervals at a decreasing rate. In flash marketing, purchase time limit is short, so we can use the difference between the responses as the basis of customer segmentation defined as "recency index" in later section.

Finally, the variety of coupons purchased also provides an important perspective. Customers who purchase coupon in only one genre easily lose their interest about the genre, so they may stop purchasing coupons from the site. However, customers who purchase coupons in various genres are more likely to retain their interest, even if they lose their interest in only one genre. To reflect this viewpoint, we apply the concept of information entropy, which is referred to as "purchase diversity".

The following sections define customer's segmentation using three indices mentioned above with corresponding analyses.

2   Indices for Customers and Coupons

In this section, we will first describe the data set, and then show the definition of each index and some preliminary results using those indices.

2.1   Overview of the Data

The data are coarsely-sampled and have some restrictions as mentioned earlier. Since fragmented data may create noise in our analysis, we have filtered this data set so that each user has at least two purchased coupons, and each coupon has at least two purchasers. Finally 7573 customers and 6646 coupons are extracted.

---

[2] New release time of this web site is almost noon.

Table 1 shows the distribution of the number of customer purchases and the number of coupon sales.

Coupon genres included in the original data set are designed for the real operation in the group buying website. Hence, we have rearranged coupon genres for the purpose of our analysis as follows : 'restaurant', 'relaxation', 'health & beauty', 'grocery & gourmet food', 'travel' and 'cosmetics'.

Table 1. The Number of Purchases by Customers and the Number of Coupon Sales

Num. of purchases (N)	Num. of customers	Num. of sales (M)	Num. of coupons
$2 \leq N < 5$	2663	$2 \leq M < 5$	3131
$5 \leq N < 10$	2439	$5 \leq M < 10$	1719
$10 \leq N < 15$	1272	$10 \leq M < 15$	731
$15 \leq N < 20$	573	$15 \leq M < 20$	344
$20 \leq N < 25$	283	$20 \leq M < 25$	209
$25 \leq N < 30$	160	$25 \leq M < 30$	109
$30 \leq N$	183	$30 \leq M$	403
total	7573	total	6646

2.2 Popularity Awareness Index (PAI) and Coupon Popularity Index (CPI)

HITS algorithm (Hyperlink-Induced Topic Search)[5,6] and PageRank algorithm [4] are well known web page ranking methods. The idea behind HITS algorithm is based on the concept of hubs and authorities in the Internet hyperlink structure. That is, a good authority page is linked by many good hub pages, and a good hub page has links to many good authority pages. HITS algorithm can rank each web page as hub site and authority site, based on the recursive definition.

We apply the HITS algorithm to explain relations between customers and coupons. In shopping behaviors, the concept of 'goodness' is not as obvious as Web pages. So we define 'popular coupons' as coupons purchased by customers who are conscious of popularity, and these customers who purchase popular coupons as 'popularity-aware customers'. Based on these recursive concepts, we can rank the 'popularity awareness' and 'coupon popularity', which are called *popularity awareness index* (PAI) and *coupon popularity index* (CPI), respectively. The details of the ranking methods are described as follows.

Let the PAI for customer i be u_i, and the CPI for coupon j be m_j. We consider the vector of PAIs : $\boldsymbol{u} = (u_1, u_2, \ldots, u_{n_u})^{\mathrm{T}}$, and the vector of CPIs : $\boldsymbol{m} = (m_1, m_2, \ldots, m_{n_m})^{\mathrm{T}}$ where n_u shows the number of customers and n_m shows the number of coupons. We can compute $\boldsymbol{u}$ and $\boldsymbol{m}$ by assigning the initial value $\boldsymbol{m}^{(0)}$ and calculating iteratively, where N denotes the number of repeat count, and $\boldsymbol{u}^{(N)} = (u_1^{(N)}, u_2^{(N)}, \ldots, u_{n_u}^{(N)})^{\mathrm{T}}$, $\boldsymbol{m}^{(N)} = (m_1^{(N)}, m_2^{(N)}, \ldots, m_{n_m}^{(N)})^{\mathrm{T}}$. In the following equations, $i \to j$ represents that customer i purchased coupon j.

$$\bar{u}_i^{(N+1)} = \sum_{j:i \to j} m_j^{(N)} \quad (i = 1, \ldots, n_u)$$

$$\overline{m}_j^{(N+1)} = \sum_{i:i \to j} u_i^{(N+1)} \quad (j = 1, \ldots, n_m)$$

$$u^{(N+1)} = \overline{u}^{(N+1)} / \|\overline{u}^{(N+1)}\|_2$$

$$m^{(N+1)} = \overline{m}^{(N+1)} / \|\overline{m}^{(N+1)}\|_2$$

$\|\cdot\|_2$ represents L_2 norm, that is, each $u^{(N+1)}$ is normalized so that square sum of each element is 1. Now let $a_{ij} (1 \leq i \leq n_u, 1 \leq j \leq n_m)$ be the purchase flag (0 or 1) by customer i for coupon j, and let

$$A = \begin{pmatrix} a_{11} & a_{12} & \cdots & a_{1n_m} \\ a_{21} & a_{22} & \cdots & a_{2n_m} \\ \vdots & \vdots & \ddots & \vdots \\ a_{n_u 1} & a_{n_u 2} & \cdots & a_{n_u n_m} \end{pmatrix},$$

then we can get

$$u^{(N+1)} = Am^{(N)}, \quad m^{(N+1)} = A^T u^{(N+1)}.$$

From the above mutual recursive equations, we can get

$$u^{(N+1)} = AA^T u^{(N)}, \quad m^{(N+1)} = A^T Am^{(N)}.$$

It is well known that $u^{(N)}$ and $m^{(N)}$ converge to the eigenvectors of AA^T and $A^T A$ respectively. These normalized eigenvectors are shown by u, m. We calculated u, m for 7573 customers and 6646 coupons based on the above framework. Moreover, we prepared PAIs for the first 6 months in order to observe the situation for the first 6 months. There are 3936 customers who have made purchases in July and August, 2011, therefore we calculated the indices based on the purchase behaviors of these customers until the end of December 2011.

We define the rank of PAIs sorted by descending order as 'popularity awareness index rank (PAI rank)'. The top-ranked customer is more aware of and sensitive to coupon popularity when compared to other customers.

2.3   Recency Index (RI)

We define the *recency index* (RI), P_i for customer i, which represents customer's purchase response. Let I_i be a set of purchased coupons by customer i, $r_k(i)$ be the order of purchase by customer i among all customers who purchased coupon $k(\in I_i)$, and M_k be the total number of purchasers for the coupon k. For example, suppose that there are 100 purchasers for a coupon k, and customer i is the 10th customer to purchase the coupon, $M_k = 100$, $r_k(i) = 10$.

We will define the recency index P_i (RI), as a normalized value of the sum of differences between the number of purchases which took place before user i and the number of purchases after user i :

$$P_i = \frac{1}{|I_i|} \sum_{k \in I_i} ((M_k - r_k(i)) - (r_k(i) - 1)),$$

In comparison, customers with negative RIs purchase coupons after other customers, while customers with positive RIs purchase coupons ahead of other customers.

As is the case with PAIs, in order to observe users' behavior during the first 6 months, we define 'RI for the first 6 months' based on the purchase behavior until December 2011, where customers have purchased in July or August 2011 (3936 persons). 'RI rank' also shows the rank of customers, sorted by descending order. The top-ranked customers in the RI make coupons purchases faster than other customers.

2.4 Purchase Diversity (PD)

Some users may only purchase coupons of a specific genre, whereas other users may purchase coupons in various genres. To quantify the diversity of purchases, we apply the concept of information entropy.

Purchase diversity (PD) can be defined using the purchase ratio for each genre as follows. Let i $(i = 1, \ldots, n$; where n is a number of customers) be a customer, and j $(j = 1, \ldots, m$; where m is a number of genres) be a genre. Thus the PD, B_i for customer i, can be defined as follows.

$$B_i = - \sum_j p_i^j \log_2 p_i^j$$

where p_i^j refers to the purchase ratio of genre j for the total purchase amount by user i. Table 2 shows the simple example of the number of purchases in each genre by each user and the PD for each user, which represents a degree of variety of purchased coupons by each customer.

Table 2. Example of PD (Purchase Diversity)

Customer	restaurant	travel	cosmetics	relaxation	PD
id1	3	2	1	2	1.906
id2	0	0	5	1	0.650
id3	10	0	0	0	0.000

In fact, if a user i only buy coupons of a specific genre, $B_i = 0$. On the other hand, if a user buy coupons in various genres, the value of PD will be high. The maximum value of PD is determined by the number of genres. In our case, if we adopt six genres, the maximum value of PD is $\log_2 6 \simeq 2.585$.

3 Experimental Results

3.1 Distribution of the Basic Indices

Popularity Awareness Index (PAI) Table 3 shows the top 10 customers with high PAIs, at which these top tier customers prefer popular coupons. For

Table 3. Top 10 Customers in PAIs

Sex	Age	PAI	restaurant	relaxation	health and beauty	grocery & gourmet food	travel	cosmetics	catalog price	cut-rate price
			Number of Purchase						Total Price (Yen)	
m	36	0.11348	2	0	0	0	1	8	81710	23025
f	36	0.09160	0	8	6	3	0	16	298125	76353
m	36	0.08995	19	0	0	1	1	7	124274	38772
f	55	0.08380	22	5	0	14	5	4	301307	131878
f	38	0.07688	1	1	0	2	1	6	115335	30422
f	52	0.07675	23	0	0	3	0	4	173451	54692
f	50	0.07412	25	1	0	0	2	4	246892	111273
f	34	0.07326	4	0	0	1	0	8	65947	22447
f	28	0.07277	22	5	2	5	3	3	412171	152693
f	42	0.07259	3	1	0	2	3	6	170490	45115

example, the top customer in this list purchased 1st, 8th, 20th, 47th, 618th coupons in CPIs.

Table 4 shows the comparison of average PAI ranks by gender and age bracket. It is noted that the PAIs of female customers are, in general, higher than male customers. This implies that middle-aged women have a higher tendency to purchase popular coupons.

Table 4. The Comparison of PAI Ranks (Average) by Gender and Age Bracket

Gender	20's	30's	40's	50's	60's
Female	4440	4071	3450	3195	3032
Male	4495	4066	3931	3695	3834

As for the prediction ability of future behaviors, we compared the behaviors for the last 6 months for group with high PAI against group with low PAI for the first 6 months (the two groups are divided based on average value). The average frequency (during the last 6 months) of high PAI group is 5.306, whereas the frequency of low PAI group is 3.201. This result shows that the PAIs are strong explanatory variables to predict future behaviors.

Coupon Popularity Index (CPI). Table 5 shows the top five coupons in CPI. The catalog price and the discount rate are shown in the table, however, there are no correlation in fact, between CPI and catalog price, or CPI and the discount rate. Thus it is concluded that the coupon popularity does not correspond to the price. It is observed that there are many cosmetics and restaurant coupons among the popular coupons. Finding out the characteristics or features for popular coupons can be the future research problem.

Table 5. Top 5 Coupons in CPIs

Genre	Item Name	Number of Sales	CPI	Discount (%)	Catalog Price (Yen)
cosmetics	Uruwoeet facial mask	463	0.584	78	7200
restaurant	Baumkuchen (Special Price)	353	0.176	50	1260
restaurant	Waffle Set (Special Price)	221	0.167	54	1100
travel	Pair Ticket with Breakfast	290	0.160	55	17000
cosmetics	White Moisture Cream	251	0.157	81	7980

Recency Index (RI). Figure 2 shows the distribution (number of customers) of RIs. RIs are distributed almost symmetrically around the center. As with PAI, we compared the behaviors during last 6 months, for group with high RI against the group with low PAI in the first 6 months (the two groups are divided based on average value). The average frequency of purchase for the high RI group during last 6 months is 4.444, whereas the frequency for low PAI group is 4.061. Although RIs are less effective explanatory variables than PAIs, RIs still contain informative features for future activities.

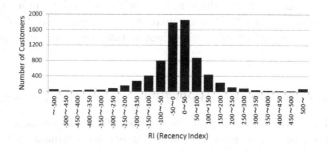

Fig. 2. Distribution of Recency Points

As shown previously, the relationship between PAIs/RIs for the first 6 months, and average purchase frequencies for the last 6 months is significant. Combining these variables enables the precise prediction of purchase frequency in the future. Table 6 shows the relationship between PAIs and RIs for the first 6 months, and average purchase frequency for the last 6 months. By combining these two parameters, the frequencies of the group with high PAIs and RIs is two times higher in the last 6 months compared to the group with lower PAIs and RIs.

Purchase Diversity (PD). Table 7 shows the comparison of numbers of purchases for last 6 months by increase/decrease of PD corresponding to the number of purchases for the first 6 months (N). Although the difference between decrease and increase is not significant when $10 < N$, there are clear differences in the

Table 6. PAIs and RIs for the First 6 Months and the Average Purchase Frequency for Last 6 Months

	RI (High)	RI (Low)
PAI (High)	5.531	5.069
PAI (Low)	3.304	3.103

case of $N \leq 5$ and $6 \leq N \leq 10$. In fact, in these two cases, the frequency of the increase group is more than 150% than the decrease group. From the results, we can conclude that purchase variety is considerably important if we expect continual activity for the last 6 months. Therefore it is recommended to promote a wide variety of coupons to the group with low purchases as marketing strategy. On the other hand, some customers in the high purchase group have solid base of coupons purchased, thus it is not necessary to promote other genre of coupons for these users.

Relationship Among PAIs, RIs and PDs. So far we have defined three kinds of parameters to characterize each customer's behavior, that is, popularity awareness index (PAI), recency index (RI) and purchase diversity (PD). Figure 3 shows the relationship among three indices, where each data point in the plot represents a customer, and "CC" is correlation coefficient. From this observation, we use these three indices as independent explanatory variables in the next subsection.

3.2 Analysis for Business Application

In this section, PAIs, RIs and PDs are analyzed from the business application viewpoint. First of all, we define active customers and dormant customers, and build a decision tree model to classify these two classes. Customers who purchased coupons in July or August 2011, with three times or more purchases made during the first 6 months are selected. Of the selected customers, customers who purchased more than three months in the last 6 months are defined as active customers (ACTIVE), and customers who did not purchase any coupon for the same period are defined as dormant customers (DORMANT). The number of active and dormant customers are 1700 and 153 respectively. The classification factors generated from the decision tree models are examined.

Table 7. Comparison of the Numbers of Purchase for Last 6 Months

	Number of Purchases (N) for First 6 Months		
	$N \leq 5$	$6 \leq N \leq 10$	$10 < N$
Decrease PD(6 month → 1 year)	2.017	4.014	8.273
Increase PD(6 month → 1 year)	3.303	6.149	9.265

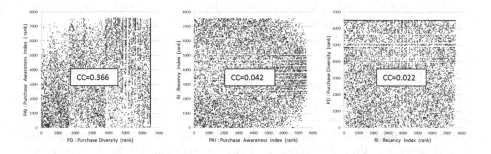

Fig. 3. Relation Among PAIs, RIs and PDs

Secondly, we are proposing a method to segment active customers with PAI rank, RI rank and PD rank. For each rank, customers are divided into two segments which are higher rank segment and lower rank segment by using the median of the ranking. Subsequently, we obtained eight customer segments. Of all segments, we focus on two segments with all higher rank or all lower rank, create classes in each segment, and build decision tree models used for classification problems.

Table 8. Explanatory Variables and Their Contents

Explanatory Variables	Contents
gender	male or female
age	The range is from 20 to 79.
NUM6	The number of total purchased coupons for the first 6 months.
restaurant	
relaxation	
health_&_beauty	The number of purchased coupons for each genre
grocery_&_gourmet_food	for the first 6 months.
cosmetics	
travel	
PAI6,PD6,and RI6	The rank of PAI, PD, and RI for the first 6 months, respectively.

In the following decision tree model, objective variables are based on the purchasing behaviors in the last 6 months. Thus the purchasing behaviors only for the first 6 months are used as explanatory variables. Table 8 denotes the explanatory variables from customer's characteristics and the purchasing behavior for the first 6 months. The decision tree models are constructed with the WEKA[3] J48 algorithm, and five folds cross validation method is used to test the model. In decision tree models shown in Figure 4, 6, 7, the description in the branch node denotes the branch condition, and the description in the leaf node denotes the branch condition, the predicted class, and the actual result for the prediction

---

[3] http://www.cs.waikato.ac.nz/ml/weka/

(the number of instances which match the rule, followed by how many of those instances are incorrectly classified, separated by "/").

Table 9 shows the evaluation index from the decision tree model for ACTIVE and DORMANT class. We can estimate the validity of class prediction using F-measure for both classes. Figure 4 shows the decision tree for active customers and dormant customers. According to the decision tree, the first division is decided by the number of coupons purchased from restaurant. The conditions of the key prediction rule (as shown in grey node (1)) of dormant customers are described as : restaurant coupons are purchased less than two times, travel coupons are purchased less than four times, and grocery and gourmet food coupons are purchased less than five times. On the other hand, the conditions of the key prediction rule of active customers are described as : restaurant coupons are purchased more than three and less than ten times, and cosmetics coupons are never purchased (as shown in grey node (2)), otherwise restaurant coupons are purchased more than nine times (as shown in grey node (3)). From the above observations, it can be noted that many of active customers purchase coupons in relation to restaurant, travel, and grocery and gourmet food categories simultaneously. Otherwise these customers are likely to purchase three or more times coupons in the restaurant category. From this result, we can conclude that customers who are satisfied with the coupon purchase in the restaurant, travel, or

Table 9. Evaluation Index of the Decision Tree Model for the Active Customer and the Dormant Customer

	Accuracy	Class	Precision	Recall	F-Measure	ROC Area
Training data	77.12%	ACTIVE	0.779	0.758	0.768	0.806
		DORMANT	0.764	0.784	0.774	0.806
Test data	63.07%	ACTIVE	0.645	0.582	0.612	0.669
		DORMANT	0.619	0.680	0.648	0.669

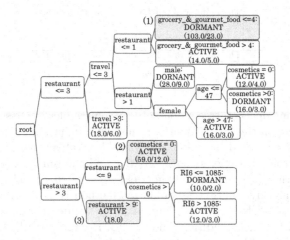

Fig. 4. Decision Tree for the Active Customer and the Dormant Customer

grocery and gourmet food categories, with repeated purchase of these coupons, are likely to become active customers.

Next, we will analyze active customers in detail. The PAI rank, RI rank and PD rank for the first 6 months and for 12 months are calculated respectively for active customers. Then, each index is divided into two segments, higher class and lower class after 6 months and after 12 months by using the median of ranking for each. The left side of Figure 5 shows eight segments after 6 months. Each box on the left side of the figure includes the values of PAI6, PD6 and RI6 (higher and lower), and the number of customers who belong to the segment. Of these segments, the segments which have all higher rank or all lower rank are called HIGH class and LOW class, respectively. Then two decision tree models were built as follows : The first one generates classification rules for the customers who belong to HIGH class. This model, which we call HIGH class model, classifies the group of higher class customers who remain in higher class (STAY), and group of lower class customers who have moved down to lower class (DOWN). The second one builds classification rules for customers who belong to the LOW class. This model, which we call LOW class model, aims to classify customers who remain in lower class (STAY) or who have moved up to higher class (UP).

Table 10. Evaluation Index of the Decision Tree Model for the Higher Class

	Accuracy	Class	Precision	Recall	F-Measure	ROC Area
Training data	81.11%	DOWN	0.680	0.836	0.750	0.870
		STAY	0.905	0.798	0.848	0.870
Test data	65.00%	DOWN	0.477	0.344	0.400	0.616
		STAY	0.706	0.807	0.753	0.616

Table 10 and Figure 6 illustrate the performance index and the decision tree of the HIGH class model, respectively. In this case, the number of customers in STAY class and DOWN class are 119 and 61, respectively. Table 10 shows the prediction for STAY class is more precise than the prediction for DOWN class for test data set.

The major rule antecedent that distinguishes STAY class holds the condition where the value of PD and PAI is higher (the PD6 and PAI6 is lower) as shown in grey node (1) of Figure 6, or when the number of relaxation coupons purchased are more than two as shown in grey node (2). On the other hand, the major rule antecedent that distinguishes DOWN class holds the condition when the number of relaxation coupon purchased is less than three, and the value of PD is low (PD6 is high), and NUM6 is less than 14 as shown in grey nodes (3) in Figure 6. From the above observations, high PD and high PAI are important classification factors to determine whether customers are likely to remain in higher class. On contrary, customers who have low diversity and purchase little coupons are likely to move down to a lower level segment. We can guess that diversity of coupons purchased and the popularity of coupons are related to user's interests in coupons. Thus, if the customer only purchases a few homogeneous coupons, it

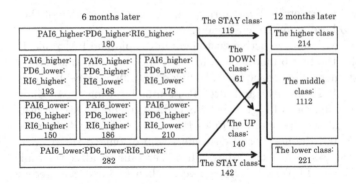

Fig. 5. Movement Among the Classes

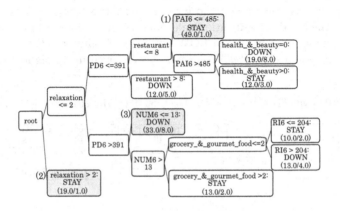

Fig. 6. Decision Tree which Classifies the Customers into STAY High Group and DOWN Group

could be a red flag that the customer will soon lose their interest in purchasing the coupon.

Similarly, Table 11 and Figure 7 illustrate the performance and the decision tree of LOW class model, respectively. From the decision tree in Figure 7, major rule antecedents that distinguish the UP class are emerged as shown grey node (1)~(3). The explanatory variables such as gender, age, and RI6 are related to the rules. On the other hand, when male customers with lower PD (PD6 is large) and lower RI (RI6 is higher) satisfied the other purchasing conditions, they are predicted as the STAY class as shown in grey node (4). From these observations, we can deduce that this group of lower class customers are relatively young, curious, and tends to make impulsive purchases. In addition, this group of customers is not interested in purchasing homogeneous coupons repeatedly.

Based on the three cases mentioned above, in order to encourage customers to make repeated purchases for a prolonged period, it is important to entice

Table 11. Evaluation Index of the Decision Tree Model for the Lower Class

	Accuracy	Class	Precision	Recall	F-Measure	ROC Area
Training data	79.43%	UP	0.820	0.750	0.784	0.847
		STAY	0.773	0.838	0.804	0.847
Test data	59.93%	UP	0.606	0.550	0.577	0.612
		STAY	0.594	0.648	0.620	0.612

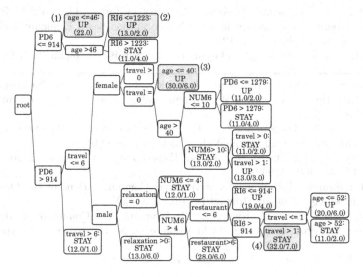

Fig. 7. Decision Tree which Classifies the Customers into STAY Lower Group and UP Group

them to purchase coupons from different categories. For higher class customers and customers moving up to a higher level, it is important to customize coupons tailored to their interests and curiosity drive. For example, in order to increase the number of active customers, the retailer should formulate a strategy to draw the customers' interest to purchase restaurant coupons for three or more times.

4 Conclusion and Future Works

In this paper, we have proposed three kinds of indices, popularity awareness index (PAI), purchase diversity (PD) and recency index (RI). By using these indices, customers can be segmented into several interesting classes, with corresponding analysis for each class. This approach is effective when the price of the product is irrelevant, which is applicable in this case. When considering other scenarios, this technique can be applied for analysis in other areas such as music CD market in which the price of the product is almost same.

As future works, the detail of movement of the 8 segments should be considered, which can be done by further dividing each index rank to reveal more detailed attributes. In some cases, it may be effective to use only two indices of the three indices. We are also interested in applying our framework on the full set of real data in other areas, to substantiate the effectiveness of the model.

References

1. The Ministry of Economy, Trade and Industry in Japan: The FY2012 Research on Infrastructure Development in Japan's Information-based Economy Society (E-Commerce Market Survey) (2012)
2. Integrated service site for group buying coupon websites in Japan (COUPON-JP): http://coupon-jp.com/
3. Nomura, S., Oyama, S., Hayamizu, T., Ishida, T.: Analysis and Improvement of HITS Algorithm for Detecting WEB Communities. Journal of Systems and Computers 35(13) (2004)
4. Brin, S., Page, L.: The Anatomy of a Large-Scale Hypertextual Web Search Engine. In: Seventh International World-Wide Web Conference (1998)
5. Kleinberg, J.: Authoritative Sources in a Hyperlinked Environment. Journal of the ACM (JACM) 46(5) (1999)
6. Uno, H.: Ranking Techniques for web pages. Communications of the Operations Research Society of Japan 57(6) (2012) (in Japanese)
7. Gibson, D., Kleinberg, J., Raghavan, P.: Inferring Web Communities from Link Topology. In: Proc. of the 9th ACM Conference on Hypertext and Hypermedia (1998)
8. Rogers, E.M.: Diffusion of Innovations. Free Press (2003)

Choice Modelling and Forecasting Demand for Alternative-Fuel Tractors

Ginevra Virginia Lombardi and Rossella Berni

[1] Dept. of Economic and Business, University of Florence, Via delle Pandette 32,
50127 Florence, Italy
[2] Dept. of Statistics, Informatics, Applications, University of Florence, Viale
Morgagni 59, 50134 Florence, Italy
berni@disia.unifi.it, gvlombardi@unifi.it

Abstract. This paper presents a study focused on potential demand for agricultural multi-functional electric tractor. In this context, the willingness-to-pay is investigated in order to establish the potential diffusion of an electrical solar tractor, by considering different levels of key attributes related to environmental, technical and economical characteristics of different version of alternative fuel tractors. The study is carried out through a choice-experiment and the application of multinomial discrete choice models, by considering heteroscedasticity of the respondent and heterogeneity across alternatives.

Keywords: Electric agricultural multi-functional vehicle, Environmental attributes, Random Utility Models, heteroscedastic extreme value model, Willingness-to-Pay.

1 Introduction

The increased concerns on climate change due to greenhouses gas (GHG) emissions in addition to air pollution effect on human health have given deep impulse to researches on electric vehicles. The researches are focused on exploring technology needs as well as market requirements for enhancing the electric vehicles acceptance and success in public opinion, [2]. The scientific efforts are frequently adressed to different aspects of developing electrical vehicle usage in urban contest, while have seldom analysed the application of the electric engine for agricultural purposes. Agriculture is acknowledged as a significant source of global greenhouse gas emissions, also if not all agriculture systems have the same implications in terms of contributions to climate change [22]. Industrial or conventional agricultural practices make use of high levels of agro-chemicals and high degrees of mechanization. These practices are made possible through increasing consumption of fossil fuels to power agricultural machinery and to support increased level of irrigation and chemical inputs. Reducing fossil fuel use in agriculture may be an objective for bolstering sustainability of industrial or conventional agriculture. A prototype of solar powered multi-functional agricultural vehicle (RAMseS: Renewable Energy Agricultural Multipurpose for Farmers) was developed in the research project financed by European Commission

P. Perner (Ed.): ICDM 2014, LNAI 8557, pp. 115–129, 2014.

under the 6th Framework Program, to offer opportunities for lessening the GHG and pollutant emission in intensive agricultural system ([19]; [20]). This paper focuses on the application of Choice Experiments and Random Utility Models (RUM) to explore which factor would facilitate the farmers' acceptance of the solar tractor and which would represent barrier to its diffusion. The farmers acceptance was not exclusively determined by costs but by a complex judgement that can be hardly evaluate before the electrical tractor have been produced and supplied by the existing market. To this end, the study is carried out by applying specific RUMs, in particular the Heteroscedastic Extreme Value (HEV) model, to explore which attributes would facilitate the farmers' acceptance of the solar tractor and which would represent barrier to its success, taking into account variability across alternatives. More precisely, preference measurements are analyzed in order to investigate the consumer preferences in term of different key attributes of the solar tractor to establish which attributes are more suitable for improving the use of solar powered electrical multi-functional tractor and which are obstacles to that. Farmers were asked to focus on different level of a set of key elements: saving achievable on fuel cost, added price premium to purchase the tractor, operational costs, battery replacement costs and environmental performances. The study was carried out in the Italian nursery plant sector, an intensive industrialised production system, located in Pistoia, a Province in the North of Tuscany. The 137 farmers were asked to give their preferences to three choice-sets, each formed by three alternatives, related to three different tractor' version (A: RAMses, electrical; B: Better, bio-fuel; C: ProGator 2030A, diesel) with different environmental, technical and economic attributes level. It must be noted that the response variable is defined as the choice of one alternative on three ones. A background questionnaire was supplied together with choice-sets to analyze farmer characteristics and to describe his/her farm typology. The paper is organized as follows: Section 2 includes a brief literature review on choice modelling and a short review on alternative-fuel demand models, Section 3 describes a general introduction to the utility framework, while Section 4 contains a brief description of the theory relating to the RUMs applied in this case study, by focusing on HEV model; Section 5 includes survey and data description useful to describe this research; the outcome of the model results and the discussion are reported in Section 6; the final remarks follow.

2   Literature Review

In this section a review on choice modelling is outlined; undoubtedly, many developments and improvements in consumer/user's preferences were achieved in the last two decades. In this brief review, we focus on recent advances, for further references on choice modelling and choice experiments see also [5]; furthermore, in subsection 2.2 a brief review on stated preferences and alternative-fuel vehicle demand models is reported.

2.1   Choice Modelling Advances

By considering the experimental design and the statistical modelling, a further and clear distinction must be made when we refer to preference measurements or, more in general, to the preference theory. Hence, we deal with Stated Preferences (SP), where we define as SP the preference of a respondent related to a hypothetical scenario represented by an alternative in a choice-set. However, in the literature, some recent developments are also reported in the Revealed Preference (RP) case, which is defined as the preference of the respondent about a real situation, such as in [25].

In the Stated Preferences context a choice-set is formed by a set of alternatives, opportunely selected from an experimental design, named choice experiment; the respondent is asked to give his/her preference within each choice-set. By considering the experimental design with its optimality criteria and the related statistical models, the class of RUM is largely applied and developed in literature, see among the others: [8], [15], [16],[17], [21], [27], [31],[33].

When considering the consumer's choice modelling, experimental designs and statistical models are closely connected [29], [34] and the properties of one design affect the corresponding model. When these properties do not exist in the design, this must be taken into account in the model. This is the case of an improvement in the design optimality specifically defined for a Mixed Multinomial Logit (MMNL), [23]; on the other hand, when considering the respondents' heterogeneity, a specific design matrix for each respondent is planned [24], by including the heterogeneity evaluation directly in the design step instead of the model step. Within the choice experiment step, optimality criteria, above all D-optimality, ad-hoc algorithms and specified information matrices for the experimental design involved were entirely defined in 1990's [35]. Further developments are related to the construction of optimal or near optimal designs with two-level attributes for binary choices in the presence of the first order interactions, [28], or when optimal designs are defined with mixed-level attributes, [10]. More recently in [7] several algorithms are compared (in draws within the Pseudo Monte-Carlo simulation method) to select efficient Bayesian designs.

Note that a common feature of recent years is to create a link among designs and models together with the need of a guiding thread between manufacturers and consumers. The paper [23] reflects the strict connection between experimental designs and statistical models, because they suggest an experimental design with ad-hoc properties for a Mixed Multinomial Logit. This model, belonging to the class of RUMs, is certainly the most widely applied and developed model in recent years for the choice experiment situation. Its success is easily explained when considering the theoretical results of [15], [21], [31]. The last developments of this model include its relationship with the latent class model, in order to create a finite number of respondent groups [8], [13], [17], [26]. Nevertheless, the HEV model may be viewed as a competitive model with respect to the Mixed Multinomial Logit to measure over-dispersion and to identify the cause and the structure of such variability.

2.2   A Short Review on Alternative-Fuel Vehicle Demand Models

Most of the early researches on alternative-fuel vehicles demand are exclusively addressed to electric vehicle. Stated preference model (SP) are widely applied to forecast demand for vehicle which are not supplied by the existing market. In [2], authors implement SP models to analyse the potential demand for hypothetical alternative-fuel vehicles, including diesel vehicle in the choices. In [18], authors underline that SP models are useful to investigate potential demand for hypothetical vehicle even though the respondent's choice may be different when considering the hypothetical market and the real market; a comparison between SP and Revealed Preferences (RP) models is conducted in [32], concluding that the two models are not providing excellent prediction although SP seems to be more reliable than RP in some circumstances.

Many researches try to integrate the SP and RP to improve forecast quality. In [30], the author analysed the market share for different version of non-gasoline-powered auto vehicles, concluding that, by the year 2000, 2.3% of passenger will be transported by electric vehicles. Other studies on alternative-fuel vehicles demand consider other alternative-fuel vehicles in addition to the electric vehicles; see among the others: [1], [9], [11]. The literature on forecasting analyses includes models for elasticity calculation, [11] or willingness to pay (WTP) model, [12].

By considering the existent literature and the preferences towards electrical vehicles, the market share is very low $(3 - 4\%)$; these results are coherent with results obtained by the case-study illustrated in Section 5.

3   A General Framework for Utility Modeling

In order to define the discrete choice models applied in this paper, we briefly introduce the fundamental elements of the utility theory [14].

As first step, the class of Random Utility Models (RUM) is defined. In general, every alternative is indicated by j $(j = 1, ..., J)$, while i denotes the consumer/user $(i = 1, ..., I)$. Each alternative will be characterized by a vector of characteristics; in what follows, price and amount of investments, farmers' characteristic and environmental practices. Thus, the following expression is characterized by a stochastic utility index U_{ij}, which may be expressed, for each unit i, as:

$$U_{ij} = V_{ij} + \epsilon_{ij} \tag{1}$$

where V_{ij} is the deterministic part of utility, while ϵ_{ij} is the random component. The random component is in general supposed independent and Gumbel or type I extreme value distributed. In the following formulas, (2) and (3), the probability density function and the cumulative distribution function (CDF) of the Gumbel distribution are defined:

$$\lambda(\frac{\epsilon_{ij}}{\theta_j}) = exp^{-\frac{\epsilon_{ij}}{\theta_j}} exp^{-exp^{-\frac{\epsilon_{ij}}{\theta_j}}} \tag{2}$$

$$\Lambda_{ij}(\frac{\epsilon_{ij}}{\theta_j}) = exp(-exp(-\epsilon_{ij}/\theta_j)) \tag{3}$$

where θ_j is the scale parameter related to the j-th alternative.

In the RUM, the individual is assumed to choose the alternative j that gives the highest level of utility, where the alternative j belongs to the choice-set C. Let the individual's indirect utility function for the alternative j be represented by:

$$U_j(q_j, y - p_j, \epsilon_j) = V_j(q_j, y - p_j) + \epsilon_j \tag{4}$$

From the researcher's perspective, the indirect utility function has two components. The first, $V_j(q_j, y-p_j)$ represents the observable portion of the individual's indirect utility function, with vector of quality characteristics q_j, income y, and price of the single product p_j. The second component of indirect utility is ϵ_j, the unobservable part of the individual's indirect utility function.

For a given choice occasion, the individual will choose the alternative j if:

$$V_j(q_j, y - p_j) + \epsilon_j \geq V_k(q_k, y - p_k) + \epsilon_k; j \in C, \forall k \in C. \tag{5}$$

Note that, just because a part of the indirect utility function is not observable, indirect utility must be expressed by:

$$v(q, y - p) = E[\max\{V_k(q_k, y - p_k) + \epsilon_k; \forall k \in C\}] \tag{6}$$

where the expectation of the right-hand side of (6) is the researcher's expectation across the random unobservable portion of the individual's utility function. Therefore, the probability of an individual i choosing the product according to the j alternative is modelled as:

$$P_i(j) = P(j|k \in C, w_i) \tag{7}$$

where w_i represents a vector of individual specific characteristic. For the purposes of the subsequent analysis we can consider the Multinomial Logit model, which can be seen as the basic model for the conditional logit described in the next section; this probability can be written as:

$$P(y_i = j) = P_i(j) = P(j|k \in C, w_i) = \frac{exp^{v_j}}{\sum_{k \in C} exp^{v_k}} \tag{8}$$

4 The Discrete Choice Models

In order to define the discrete choice models applied and discussed in this paper, we briefly introduce the fundamental elements of the related theory; for further details see the previous cited references (Section 2.1). The class of RUM, which aims to achieve the utility maximization for the consumer, enlarges the characteristics of the Logit and Multinomial models where the IIA property is hypothesized. The relaxation of this assumption [31] is a relevant improvement because the IIA means that the choosing probability in one choice-set is independent of the presence of other attribute values or any other alternative; on the other hand, we may say that IIA derives from the hypothesis of independence and homoscedasticity of the error terms. In addition, this can also be interpreted by considering the cross-elasticity term. In fact, IIA implies an equal proportional substitution between alternatives, [31].

Furthermore, the Logit and Multinomial models do not allow to evaluate a different behavior of the consumer; i.e. each respondent, with different baseline characteristics, is treated in a similar way (the same estimate values of attributes) according only to his/her judgement.

In order to deal with the above issues, the statistical analysis is carried out through three discrete choice models belonging to this class, and, in particular, through the Multinomial Logit Model (MNL), the multinomial mixed logit and the HEV model [6].

The Multinomial Logit Model (MNL) may be also view as conditional logit model; the term "conditional" highlights that the unit i chooses the alternative j, which belongs to a set of alternatives called choice-set C_i and then the model applied is called Conditional Logit (CL). Thus, the probability of the unit i to choose the alternative j is defined as:

$$P(y_i = j) = P_{ij} = \frac{exp(x'_{ij}\beta)}{\sum_{k \in C_i} exp(x'_{ik}\beta)} \tag{9}$$

where x_{ij} denotes the value of the attribute for the alternative j and the unit i. Note that, the difference is expressed through the J values of the random variable y, which indicates the choice made from the unit i. The CL model is the basic discrete choice model applied in this paper and we remark that this model assumes the IIA property; in addition, in this case, the error term is distributed according to formula (3) without the evaluation of the scale parameter θ_j, i.e. the error terms are supposed identically distributed.

When a Mixed MNL model is considered, the general expression for a RUM model becomes:

$$U_{ij} = V_{ij} + \psi_{ij} + \epsilon_{ij} \tag{10}$$

The main feature of the Mixed MNL model, or of the Mixed logit model when the choice is binary, is the possibility to assume a general continuous distribution for the ψ_{ij} called also mixing term. In fact, a density for ψ_{ij} is defined as in the following:

$$g(\psi \mid \Phi) \tag{11}$$

where the space parameter Φ contains the fixed parameters of the distribution, such as Normal, Uniform, Log-Normal. If ψ is not evaluated, then the mixed logit reduces to the simple conditional logit; in general, the unconditional probability is equal to:

$$P(y_i = j) = P_i(j) = \int_{\psi} L_i(j \mid \psi_{ij})g(\psi_{ij} \mid \phi)d\psi_{ij} \tag{12}$$

$$L_i(j \mid \psi_{ij}) = \frac{exp(x'_{ij}\beta + \psi_{ij})}{\sum_{k \in C_i} exp(x'_{ik}\beta + \psi_{ik})}$$

Note that the unconditional choice probability $P_i(j)$ is the integral of the conditional probability of the logit model integrated over the distribution of $\psi_{ij}, \forall i, j$

and weighted according to the fixed parameters of the mixing term. Therefore, the mixed logit model allows to treat the heterogeneity of respondents through the random parameters associated to a specific attribute of an alternative. Nevertheless, the error term across alternative in not weighted, as in the following model.

The Heteroscedastic Extreme Value (HEV) model [6] is the third discrete choice model considered in this paper and belongs to the RUM class as defined in formula (1). The main feature of this model, which differentiates it by the CL model and the Mixed Logit, concerns the modified assumptions on the random component. In this model, the random component, supposed distributed as a type I extreme value distribution, formula (3), is assumed independently but not identically distributed. This different hypothesis on the random component allows us to treat differently the relaxation on the IIA property with respect to the Mixed Logit model, because, in the HEV model, the homoscedasticity hypothesis of the error terms is not assumed and, therefore, different scale parameters across alternatives are estimated. This last consideration implies that cross-elasticities are not supposed to be all equals, as in the MNL and the logit models.

The main evident advantage is that the scale parameters may be defined as the weights in order to measure the uncertainty related to the alternatives and to the attributed there involved. Furthermore, the presence of large variances for the error terms influences the effects of changing of systematic utility for the generical alternative j.

Therefore, the probability for a respondent i to choose the alternative j from a choice-set C_i is:

$$P(y_i = j) = P_i(j) = \int_\epsilon \prod_{k \in C_i; k \neq j} \Lambda \{ \frac{x'_{ij}\beta - x'_{ik}\beta + \epsilon_{ij}}{\theta_k} \} \frac{1}{\theta_j} \lambda(\frac{\epsilon_{ij}}{\theta_j}) d\epsilon_{ij} \quad (13)$$

where θ_j is the scale parameter for the j alternative and $\lambda(\cdot)$ is the probability density function of the Gumbel distribution, as in formula (2); the term $x'_{(\cdot)}\beta$ denotes the deterministic part of utility of formula (1). Note that the integral function is defined on the domain $[-\infty, +\infty]$ of the random component ϵ related to the unit i and the alternative j.

The theoretical framework of these three discrete choice models allows us to outline useful comparisons when evaluating the farmers preferences. Furthermore, the CL model is seen as the basic and simple model which does not take care of respondent's heterogeneity due to baseline variables (such as age of respondent); thus, heterogeneity is modelled in the Mixed Logit through the mixing term, $g(\psi \mid \Phi)$, where the expressed preference of respondent i, $(L_i(j))$, is measured conditioning to the personnel characteristics (ψ_{ij}).

The HEV model is considered as a further and different improvement to the CL model with respect to the Mixed Logit model. In this case the farmer preferences of respondent i are evaluated by considering a scaling term θ_j for the alternative j in the choice-set C_i, i.e. the heteroscedasticity of the error term.

It's not straightforward matter to say that the HEV and the Mixed Logit models could be considered as competitive models in order to identify and to measure the presence of an over-dispersion when modeling the consumer preferences, with respect to the CL model.

In what follows, the farmer preferences are evaluated by considering respondent's heterogeneity or the heteroscedasticity of the alternatives.

The discrete choice models are evaluated through the following goodness-of-fit criteria: the maximum gradient element, the number of iterations to reach convergence, the Likelihood Ratio (LR), the Akaike's index (AIC) and the McFadden's LR index (McFadden LRI), bounded in $[0, 1]$, which is defined as the complementary to one of the LR.

5   Data and Variables Description

The case-study involved 137 plant nursery farms [4], located in the province of Pistoia (Italy); farmers were asked to give their preferences regarding three choice-sets (N=411 stated preferences), each formed by three alternatives, relating to three vehicles (A: RAMses, electrical; B: Better, bio-fuel; C: ProGator 2030A, diesel). It must be noted that three different situations were analysed in order to assess the probability of choosing from among different vehicles, RAMses and other two tractors supplied by the real market, with particular focus on the environmental, technical and economic characteristic of the electrical one. Each situation corresponds to a single choice-set formed by alternative versions of the three vehicles. The choices are defined by considering realistic baseline technical and economical attribute for biofuel and gasoline tractors while for RAMses, a prototype not yet available on the market, hypotetical versions with different levels of the key attribute have been considered. It is pointed out that the response variable is defined as the choice of one of the three alternatives.

The attributes (Table 1), each at three levels, involved in the experiment are: Price- purchasing price of the vehicle, Cost- monthly cost or operating cost, power, emissions and noise level. The background questionnaire was composed by three main sections. The first part includes questions to explore demographic and socioeconomic characteristics of the farmer/respondent; the second part is dedicated to environmental attitude of the farmer; this section include stated preferences to forecast respondent's attitude toward the electrical tractor purchase.

Table 1. Attributes description

Attribute	levels	range
Price (euro)	19,700;35,000;40,000	[19,000-40,000]
Cost (euro)	108,00;280,00;357,00	[100,00-360,00]
Power (KW)	12.00;17.70;66.00	[12.00-66.00]
Emissions (Kg/h)	0;3.60;7.20	[0-7.20]
Noise	low;medium;high	

In the third part respondents were asked to provide detailed information about the farm structure to investigate farm typology and characteristics.

With respect to the farm's characteristics and the background questionnaire, we considered: age of farmer/respondent (Table 2); farm size (Table 3); Q54 - amount of farm machinery investments (Table 4); Q21- family-run farm (Table 5); Q24- farm equipped with electrical vehicles for people transportation (Table 6); Q61- Acceptance: the stated interest in purchasing a multi-functional electric vehicle recharged by a photovoltaic system (PV) (RAMses prototype) (Table 7); Q42-relates to farms adopting good environmental practices (Table 8).

Table 2. Distribution of farmers by age; missing values:16

Years (in class)	freq.(n)	%
≤ 40	36	29.75
41 ⊣ 55	56	46.28
> 55	29	23.97
Total	121	100.00

Table 3. Distribution of farms by size; missing values:10

Farm size (class in hectare)	freq.(n)	%
≤ 1	5	29.75
1 ⊣ 4	64	46.28
> 4	58	23.97
Total	127	100.00

Table 4. Q54-Distribution of farms by investments; missing values:3

Inv. (class in euro)	freq.(n)	%
≤ 100, 000	50	37.31
100, 000 ⊣ 500, 000	79	58.96
> 500, 000	5	3.73
Total	134	100.00

Two constants are created in order to analyse the choice preferences between: i) RAMses and bio-fuel (const-B); ii) RAMses and diesel (const-C).

It must be noted that the evaluation of constants includes the natural differences between vehicles; in fact, when comparing RAMses and bio-fuel, as well as RAMses and diesel, the differences in fuel and range autonomy are implicit. In addition, the related dummies are computed for each explicative variable; for example, by considering the farm size and the amount of farm machinery investments, three classes and six dummies are created. More specifically, when considering each of the three classes of the variable investment, two dummies are

Table 5. Q21-Family-run farm; missing values:0

Family-run	freq.(n)	%
Yes	110	80.29
No	27	19.71
Total	137	100.00

Table 6. Q24-Presence/absence on the farm of electrical vehicles for transporting people; missing values:0

El. vehicle	freq.(n)	%
Presence	125	8.76
Absence	12	91.24
Total	137	100.00

Table 7. Q61-Interest in purchasing an electrical vehicle; missing values:0

Interest	freq.(n)	%
Yes	81	59.12
No	56	40.88
Total	137	100.00

Table 8. Q42-adopting good environmental practices; missing values:0

Env. pract.	freq.(n)	%
Yes	96	70.07
No	41	29.93
Total	137	100.00

created (namely, investment-B and investment-C for the class $< 100,000 euro$), where the suffices B and C have the same meaning as with the constants. The statistical analysis was started by considering all the previously mentioned variables and attributes and their potential associations.

The statistical analysis has begun by evaluating firstly the conditional logit model, further heteroscedasticity of alternatives and heterogeneity of respondents are taken into account through the HEV and Mixed models, respectively. Nevertheless, statistical results do not reveal the presence of heterogeneity with respect to farms' characteristics which may have a potential effect on purchasing electrical vehicle, e.g RAMses; on the contrary, as detailed in the following (Section 6), the model results show a significant presence of heteroscedasticity across alternatives, due to the difficulties of respondent to choose between RAMses and the bio-diesel tractor or between RAMses and the diesel tractor.

Therefore, in the next section, we show the more interesting results obtained by applying the HEV model.

6   Model Results

For each estimated model, the most relevant results are reported by considering the estimates of coefficients, with standard errors and p-values. The correlation matrix of parameter estimates is always evaluated; values are reported when relevant for the discussion.

The results of the choice-experiment have been analyzed by considering also the results of the background questionnaire in order to forecast the behaviour of potential consumer classes based on individual characteristics and on the farm typology.

The first estimated HEV model, illustrated in Table 9, includes: const-B and const-C, Q54 (by conditioning to the class of farm investment machinery below 100,000 euro), Q42 referring to farms adopting good environmental practises and the operating cost level(monthly cost variable) including battery replacement. The const-B coefficient shows a propensity to purchase the bio-fuel version of the tractor, and the const-C coefficient shows greater propensity towards the electric tractor. In both constants the utility towards electrical tractor is positively influenced by the coefficients related to Q42 and negatively influenced by Q54. The estimated coefficient for the monthly cost shows a decreasing utility for the electrical vehicle as the cost increases in function of technology sets. In Table 10 the results relating to the second estimated HEV model are reported; variables and attributes involved therein are: const-B and const-C, Q21 (family farms), Q54 and the purchase price of the tractor. The estimated coefficients relating to const-B and const-C show a propensity towards bio-fuel tractor in const-B and electric tractor in const-C. Coefficients estimated for the family farms (Q21) report great propensity for buying electric tractor; while the estimated coefficients for the farms with the lowest machinery investment level show a propensity for the bio-fuel and the diesel version of the tractor, Q54-B and Q54-C respectively. The purchase price is negatively correlated with RAMses: as the price increases the utility of electric tractor decreases. The propensity toward electrical tractor utility is positively influenced by the respondents electric vehicle acceptance as revealed by question Q21 and Q42, and by the highest monthly cost differences between tractor alternatives. These cost differences are determined by hypothetical differences in electric tractor technology (battery cost and the battery life). Nevertheless, the electrical tractor utility is negatively influenced by the price and by the lowest monthly cost difference among alternatives. The model forecasts the propensity of purchasing tractors by considering tractor technical and environmental characteristics, price and then fuel type. The price sensitiveness is highly influenced by the tractor technology version. The results show also there is respondent's propensity to select the tractor version with higher level of both environmental and technical characteristics, preferences are in fact accorded to the alternative with the stronger engine power associated with the higher level of environmental attribute. The choice experiment shows that for the electrical version the premium price ranges from 1,000 to 5,000 euro; no WTP for premium price of 15,000 euro is accounted for electrical tractor. As to monthly cost, WTP shows a range from 250.00 to 290.00 euro per month, while no WTP is

Table 9. HEV model results: monthly cost and environmental characteristics

Coefficient	estimate	std.err.	p-value
const-B	6.678	1.081	<.0001
const-C	-2.138	1.820	0.2401
Q42-B	-0.941	0.684	0.1693
Q42-C	-2.094	0.509	<.0001
Q54-B	0.586	0.374	0.1175
Q54-C	0.794	0.419	0.0579
Cost	-0.032	0.005	0.0420
Scale-3	1.112	0.547	0.0420

Table 10. HEV model results: price of vehicle and environmental characteristics

Coefficient	estimate	std.err.	p-value
const-B	8.874	1.447	<.0001
const-C	-1.276	0.668	0.0561
Q21-B	-1.190	0.374	0.0015
Q21-C	-0.599	0.426	0.1589
Q54-B	0.641	0.359	0.0740
Q54-C	0.739	0.386	0.0557
Price	-4.5e-4	7.9e-5	<.0001
Scale-3	1.382	0.803	0.0853

revealed when the monthly cost reaches 350.00 euro per month. The WTP for enhancing technology sets in relation to power of the engine ranges from 55.00 to a maximum of 211.00 euro per Horsepower. The results obtained through the application of Mixed Multinomial Logit models do not reveal a presence of significant over-dispersion due to respondents' heterogeneity with respect to farms' characteristics influencing the propensity towards RAMses. This result may also be explained by considering the investigated population, which is composed by all the plant nursery farms located in a small geographical area, with very similar characteristics. Therefore, by considering the two general sources of variability, e.g. respondents and/or alternatives (Section 4), this study reveals a significant variability during the decision process, when respondents must express their preference and therefore when they must decide among the alternatives (vehicles). In fact, as reported in Table 1 and Table 2, the applied HEV models show significant scale-3 parameters; a further confirmation of the higher variability when respondents have to express a choice between the RAMses and diesel vehicles.

7   Final Remarks

In this paper, choice experiments and multinomial choice models are applied in order to evaluate the preferences of farmers towards a renewable-energy powered tractor. This preliminary work shows that there is a potential demand for electric tractor in agriculture and that the problem of the electric agricultural machinery diffusion in agriculture doesn't meet the concern about recharging infrastructure. The statistical analysis shows that there is farmers' general propensity toward environmental attribute of tractors and also acceptance and reliability for electric powered tractors. The farms with high machinery investement, the farm with low environmental impact and the farm equipped with electric vehicle for people transportation show higher attitude toward the electrical tractor. On the contrary, the family-run farms are more unwilling in purchasing the electrical version of the tractor. According to consumer survey literature, the results pointed out that the price of purchasing electrical machinery is the principal barrier to its diffusion. Electric tractor current technology (low power mainly due to battery efficiency) and the operating costs (determined by the battery cost and by its short life) are the main limits to the potential demand of the electric tractor in the nursery plant sector. Therefore the analysis underlines that there is a technology innovation need in order to allow the battery cost to fall and to increase battery life and efficiency (high power battery design). These steps are necessary to enhance the performances of the agricultural electrical tractor and to help raise its competitiveness in the market, given that battery costs have influence on tractor price and on its operating cost. It is relevant to policy makers that the diffusion of the electric machinery in agriculture, currently, asks for the implementation of supporting policy measures including price incentives in order to improve affordability of electric tractor for farmers. It must be noted also that there is a potential conflict in EU Common Agricultural Policy between fossil fuel subsidies and policy to support electric tractor diffusion in agriculture since the WTP for electric vehicle is higher when the fossil fuel price is high.

References

1. Alvarez-Daziano, R., Bolduc, D.: Canadian consumers' perceptual and attitudinal responses towards green automobile technologies: an application of hybrid choice models. In: 2009 EAERE-FEEM-VIU European Summer School in Resources and Environmental Economics: Economics, Transport and Environment, Venice International University (2009)
2. Beggs, S., Cardell, S., Hausman, J.: Assessing the potential demand for electric cars. Journal of Econometrics 16, 1–19 (1981)
3. Bernal-Agustin, J.L., Dufo-Lopez, R.: Economical and environmental analysis of grid connected photovoltaic systems in Spain. Renew. Energ. 31, 1107–1128 (2006)
4. Berni, R., Lombardi, G.V.: Agricultural multi-functional vehicles and the environment: choice experiments and random utility models for investigating renewable energies. Statistica Applicata-Italian Journal of Applied Statistics 22, 363–374 (2012)
5. Berni, R., Rivello, R.: Choices and conjoint analysis: critical aspects and recent developments. In: Bini, M., Monari, P., Piccolo, D., Salmaso, L. (eds.) Statistical Methods and Models for the Evaluation of Educational Services and Products' Quality. Contribution to Statistics, pp. 119–137. Springer (2009)
6. Bhat, C.R.: A heteroschedstic extreme value model of intercity travel mode choice. Transport. Res. B-Meth. 29, 471–483 (1995)

7. Bliemer, M., Rose, J., Hess, S.: Approximation of bayesian efficiency in experimental choice designs. Journal of Choice Modelling 1, 98–127 (2008)
8. Boxall, P.C., Adamowicz, W.L.: Understanding heterogeneous preferences in random utility models: a latent class approach. Environmental and Resource Economics 23, 421–446 (2002)
9. Brownstone, D., Bunch, D.S., Golob, T.F., Ren, W.: A transactions choice model for forecasting demand for alternative-fuel vehicles. Research in Transportation Economics 4, 87–129 (1996)
10. Burgess, L., Street, D.J.: Optimal designs for choice experiments with asymmetric attributes. Journal of Statistical Planning and Inference 134, 288–301 (2005)
11. Dagsvik, J.K., Wennemo, T., Wetterwald, D.G., Aaberge, R.: Potential demand for alternative fuel vehicles. Transportation Research Part B 36, 361–384 (2002)
12. Daziano, R.A., Bolduc, D.: Incorporating pro-environmental preferences towards green automobile technologies through a Bayesian hybrid choice model. Transportmetrica A: Transport Science 9, 74–106 (2013)
13. Greene, W.H., Hensher, D.A.: A latent class model for discrete choice analysis: contrasts with mixed logit. Transport. Res. B-Meth. 37, 681–698 (2003)
14. Haab, T.C., Hicks, R.L.: Accounting for choice set endogeneity in random utility models of recreation demand. Journal of Environmental Economics and Management 34, 127–147 (1997)
15. Hensher, D.A., Greene, W.H.: The Mixed Logit model: the state of practice. Transportation 30, 133–176 (2003)
16. Hole, A.: A discrete choice model with endogeneous attribute attendance. Economics Letters 110, 203–205 (2011)
17. Hynes, S., Hanley, N., Scarpa, R.: Effects on welfare measures of alternative means of accounting for preference heterogeneity in recreational demand models. American Journal of Agricultural Economics 901, 1011–1027 (2008)
18. Mannering, F.L., Train, K.: Recent directions in automobile demand modeling. Transportation Research Part B 19, 265–274 (1985)
19. Mousazadeh, H., Keyhani, A., Mobli, H., Bardi, U., Lombardi, G.V., El-Asmar, T.: Environmental assessment of RAMseS multipurpose electric vehicle compared to a conventional combustion engine vehicle. J. Clean. Prod. 17, 781–790 (2009)
20. Mousazadeh, H., Keyhani, A., Mobli, H., Bardi, U., Lombardi, G.V., El-Asmar, T.: Technical and economical assessment of a multipurpose electric vehicle for farmers. J. Clean. Prod. 17, 1556–1562 (2009)
21. McFadden, D., Train, K.: Mixed MNL for discrete response. J. Appl. Econom. 15, 447–450 (2000)
22. OECD, Climate Change and Agriculture: Impacts, Adaptation and Mitigation, Paris (2010)
23. Sandor, Z., Wedel, M.: Profile construction in experimental choice designs for mixed logit models. Marketing Science 21, 445–475 (2002)
24. Sandor, Z., Wedel, M.: Heterogeneous conjoint choice designs. Journal of Marketing Research 42, 210–218 (2005)
25. Scarpa, R., Ruto, E.S.K., Kristjanson, P., Radeny, M., Druker, A.G., Rege, J.E.O.: Valuing indigenous cattle breeds in Kenya: an empirical comparison of stated and revelaed preference value estimates. Ecological Economics 45, 409–426 (2003)
26. Scarpa, R., Thiene, M.: Destination choice models for rock climbing in the Northeastern Alps: a latent-class approach based on intensity of preferences. Land Economics 81, 426–444 (2005)

27. Scarpa, R., Thiene, M., Train, K.: Utility in WTP space: a tool to address confounding random scale effects in destination choice to the Alps. Working Paper in Economics. Dept. of Economics, Univ. of Waikato, New Zealand, vol. 15, pp. 1–22 (2007)
28. Street, D.J., Burgess, L.: Optimal and near-optimal pairs for the estimation of effects in 2-level choice experiments. Journal of Statistical Planning and Inference 118, 185–199 (2004)
29. Toubia, O., Hauser, J.R.: On managerially efficient experimental designs. Marketing Science 26, 851–858 (2007)
30. Train, K.E.: The potential market for non-gasoline-powered automobiles. Transportation Research Part A 14, 405–414 (1980)
31. Train, K.E.: Recreation demand models with taste differences over people. Land Econ. 74, 230–239 (1998)
32. Wardman, M.: A Comparison of revealed preference and stated preference models of travel behaviour. Journal of Transport Economics and Policy 22, 71–92 (1988)
33. Wen, C., Koppelman, F.: A Generalized Nested Logit model. Transportation Research: Part B: Methodological 35, 627–641 (2001)
34. Yu, J., Goos, P., Vandebroek, M.L.: Efficient conjoint choice designs in the presence of respondent heterogeneity. Marketing Science 28, 122–135 (2009)
35. Zwerina, K., Huber, J., Kuhfeld, W.F.: A general method for constructing efficient choice designs. Working paper-Fuqua School of Business-Duke University, Durham, vol. 27708, pp. 121–139 (1996)

Data Mining Cultural Aspects
of Social Media Marketing

Ronald Hochreiter and Christoph Waldhauser

Institute for Statistics and Mathematics
WU Vienna University of Economics and Business, Austria
first.last@wu.ac.at

Abstract. or marketing to function in a globalized world it must respect
a diverse set of local cultures. With marketing efforts extending to social
media platforms, the crossing of cultural boundaries can happen in an
instant. In this paper we examine how culture influences the popularity of
marketing messages in social media platforms. Text mining, automated
translation and sentiment analysis contribute largely to our research.
From our analysis of 400 posts on the localized Google+ pages of German
car brands in Germany and the US, we conclude that posting time and
emotions are important predictors for reshare counts.

1 Introduction

To a large part, marketing can be summarized as giving consumers what they
want [1, 2] . In the right contexts, a proven method is to do this informally
[3]. While informal, heard it through the grapevine communication channels
have always been important to marketing niche products and in conquering new
markets, social media enables marketers to use word of mouth propagation for
more established mainstream products as well [4, 5]. Since word of mouth is
a very powerful vehicle to transport marketing messages [6], marketers seek to
harness the power of social media to enlist users not only as consumers but as
propagators and endorsers of products [7], to e.g. make them partners in the
co-creation process of a brand. As Kitchen [5] demonstrates, social media can
be used very cost effectively, enabling marketers to reach millions of users with
only a negligible amount of resources.

It is therefore tempting to use social media marketing efforts to spread across
traditional borders and reach for new markets. However, there is a risk associated
with this: communication needs to be careful when crossing borders venturing
into the realms of other cultures.

A lot has been written about the need for culturally-aware communication
and the management of global brands in a globalized economy [8–11]. It is generally recommended [12], to develop culturally similar markets when moving
one's brand abroad. This is, because culturally accurate (i.e. functioning) translations of marketing and branding messages are very complex to produce. In this
paper we seek to examine how the access to word of mouth propagation changes

P. Perner (Ed.): ICDM 2014, LNAI 8557, pp. 130–143, 2014.
© Springer International Publishing Switzerland 2014

across cultures. To answer this question we employed data and sentiment mining techniques to the social media posts of two brands (BMW and Audi) in two countries (Germany and the US) and compared the factors that contributed to these posts being endorsed by users.

This paper is organized as follows. We will first review the literature on social media, our target platform Google+ and how marketing is done there. We then turn our attention to cultural aspects of communication and marketing. We complete this paper's theoretical part with a concise review of sentiment mining methodology. We then present the method behind our data harvest/generation and introduce the statistical models we optimized. Finally, we discuss our findings and close with some concluding remarks that point to further research.

2   Social Media Marketing

In this section we focus on a rather novel arena of marketing: social media marketing. To this end we will give a short run-down on social media and then introduce a recently becoming increasingly popular social media platform: Google+ [13, 14]. Because of its comparably young age, Google+ has not yet received widespread attention in academia, apart from its technical aspects [15]. This is perhaps due to its somewhat differing implementation of classical social media.

Social media and online social networks refer to a rather new phenomenon in human, internet-based communication. Diverging from a dogma that had been valid for about two decades, users themselves started out using blogs to regularly provide content themselves. While blogs were and still are appealing to users interested in writing longer texts, social media as a mass phenomenon took off only after the introduction of communities centered around profiles, frequent status updates and shared content creation. Today, leading examples of social media platforms are Facebook, Twitter, vKontakte and Google+, to name but a few. While the implementation details differ from a technical point of view, there is a common theoretical framework.

In the center of social media are user following relationships. They can be thought of as (directed) graphs linking up users. From a network theoretical point of view, this graph has small world properties and has a node distribution that follows a power law [16, 17], thus being very similar to actual, offline human behavior. This relationship is called following and implies that messages sent by a user will be pushed into the stream of news all of her followers receive. While some social media platforms require reciprocal fellowship relations (e.g. Facebook, LinkedIn), others don't (e.g. Google+).

Once a message gets pushed to a user, that user can decide on how to further treat the message. Besides the obvious ignoring, a user has two levels of endorsement to choose from: liking and resharing. Endorsements are then pushed further downstream the user's network. The lesser form of endorsement, liking, does usually not contain the endorsed message but only the fact that a message from the original author was endorsed and a link to that message. User interfaces will also show likes not as prominently as reshares. With reshares the entire original message gets pushed into the streams of the user's followers, just as if the user had posted the message herself.

Another corner stone of social media are profile pages. There individuals and companies can maintain a presence with information related to them [18]. Most platforms distinguish between profiles for humans and pages for businesses and brands. However, conceptually, they are the same.

Google's implementation of a social media platform, termed Google+ and albeit being quite young, is increasing rapidly in popularity [13, 14, 18]. Here, following relationships are organized in circles, that act like address books or friend lists and allow for a more targeted resharing. While implementing a lesser form of endorsement (+1ing), it is not quite clear how that endorsement functions. For one, it is used in Google's main business of information retrieval by allowing users to discover search results that have been endorsed by the people they follow. +1s also contribute to Google internal popularity metrics, i.e. recommending the circling of possible contacts based on similar +1ing behavior. Finally, +1ed posts *might* show up in a user's stream, if Google deems them algorithmically interesting. However, the exact mechanics have not yet been published by Google.

Using social media for marketing purposes is an obvious choice: being able to interact with consumers in channels usually associated with friends and interesting people we follow, makes for a very attractive marketing context. This turns marketing presence into consumer activation [5]. A priceless achievement. Added to this is the possibility of marketing messages being reshared, thus profiting from traditional offline word-of-mouth benefits [19–21].

Under these terms, marketers must aim to maximize the resharing of their messages. It is useful to think of the individual user as a filter through which a message must pass in order to reach further into the network of users [22–24]. A lot of research effort has been put into finding predictors for the expected reshare count of a message [25, 26]. Message sentiment is identified by [27, 28] as crucial predictors for resharing counts as is message length [28]. Temporal proximity and time of day are factors named in [29]'s contribution.

3   Cultural Aspects of Marketing

Starting with Geert Hofstede's massive and groundbreaking survey of intercultural communication [30], cultural aspects quickly became important when adapting marketing messages to local audiences of consumers [31, 32].

Applying Hofstede's insights to marketing, [4, 5, 12] come to the conclusion that marketing messages must be adapted to the cultural expectations of consumers for them to function. When looking at marketing in social media, this should also be true. Consumers organized into localized brand pages should differ in their preferences on marketing messages, just as they would differ in offline communication. From Hofstede's difference matrix, it is to be expected to find different marketing message preferences in the dimensions of *Individuality* and *Uncertainty avoidance* for our cases of Germany and the US. These differences

Table 1. Scores in Hofstede's cultural dimensions for Germany and the US [33]

Dimension	Germany	US
Power distance	35	40
Individuality	67	91
Masculinity	66	62
Uncertainty avoidance	65	46

should express themselves in terms of the importance of the filtering criteria per country, as introduced above.

In the next section we will review how sentiment mining can help in providing message characteristics for discriminating along the lines of cultural preferences.

3.1 Sentiment Mining

Sentiment analysis and opinion mining are sub-fields of the area of text mining, see e.g. [34]. It is a classification task and represents the computational study of sentiments, subjectivity, appraisal, and emotions expressed in text. Companies usually spend huge amounts of money to find consumer opinions using consultants, surveys, and focus groups. A cleverly implemented sentiment mining tool supports such companies to save money.

An opinion is simply a positive or negative sentiment, view, attitude, emotion, or appraisal about an entity or an aspect of the entity [35] from an opinion holder [36]. The sentiment orientation (sentiment polarity) of an opinion can be positive, negative, or neutral (no opinion).

Besides in the (sentiment) polarity of a respective Google+ post, we are interested in the emotionality of a posting. However there is no agreed set of basic emotions of people among researchers. Parrott [37] identifies six main emotions, i.e. love, joy, surprise, anger, sadness, and fear, whereby the strengths of opinions and sentiments are sometimes related to certain emotions, e.g., joy, anger. We apply the R package `sentiment` [38] to compute the following emotions: anger, disgust, fear, joy, sadness, and surprise.

3.2 Data and Methods

The generation of the data set used in this analysis started out with the identification and designation of Google+ company pages. We settled for two German companies in the automotive industry, BMW and Audi, as German car brands in different countries make a good middle ground between niche and mainstream markets [39].

Measuring culture of a company or any individual can be elusive. It is possible to use proxies like country of residence or nationality to locate an individual culturally [40]. We are following a similar path by looking at the companies'

respective local Google+ pages. We therefore chose the localized variants of the
Google+ pages of Audi and BMW for Germany and the US.[1]

The following Table 2 shows the popularities of the respective companies'
local Google+ pages. Google offers two metrics to measure the popularity of a
page. One is the circle count, i.e. the number of people that have subscribed to
push updates of that page. The other one is the +1 count that aggregates the
circle count with people +1ing the page or interacting with it in other ways.[2]

Table 2. Popularity of local G+ pages of German car brands. All data was retrieved
on January 8th, 2014.

Brand	Country	Page ID	Circle count	+1 count
Audi	Germany	audide	67486	94337
	US	AudiUSA	1623374	2116081
BMW	Germany	BMWDeutschland	84429	113853
	US	BMWUSA	39946	84042

In order to build up a data base of marketing messages, the last 100 posts
of each of these pages were retrieved. Due to different posting frequencies, the
extent of the data's retrospection varies. Table 3 summarizes these differences.
From these figures it is obvious, that English language content for the US versions
of the G+ presences is provided far more frequently than for their German
counterparts.

Table 3. Average posting frequencies and extents of the retrospection for harvested
data

Brand	Country	Posts/day	Start date
Audi	Germany	0.3112	2013-2-20
	US	0.7133	2013-8-20
BMW	Germany	0.5124	2013-6-26
	US	0.763	2013-8-29

In the following we will describe the data set in greater detail. Table 4 gives
an overview of directly measured message properties. The variable *Age* describes
how many days in the past the message had been posted. This obviously influences the number of reshares a message received or could receive. Message length
is a property that in the past has often been used successfully to predict message
reshare counts [26].

---

[1] Even though Audi and BMW are both German brands, their main Google+ pages
are international fronts. So both localized version are comparable in catering to local
audiences.

[2] C.f. http://googleblog.blogspot.co.at/2011/12/google-few-big-
improvements-before-new.html.

Table 4. Ranges, means and standard deviations of recorded message properties

Variable	Min	Mean	Max	SD
Age	3.45	98.70	325.33	70.59
Number of reshares	0.00	11.21	101.00	13.38
Number of comments	0.00	12.23	173.00	18.86
Number of +1s	12.00	170.75	928.00	150.77
Message length	0.00	262.88	1748.00	388.23

As mentioned above, another aspect that – potentially – influences the frequency with which a message is being reshared is the time of day of the original post and the day in the week. For this analysis we recorded the date and time of the original post (and converted the UTC timezone reported by Google to EST for the US and CET for Germany) and distilled the day of week from it as well as the discrete factorization of the time of day part that is given in Table 5.

Table 5. Distribution of time of day periods

Period	Frequency
6 a.m. – 9 a.m.	3
9 a.m. – 12 p.m.	53
12 p.m. – 5 p.m.	240
5 p.m. – 8 p.m.	63
8 p.m. – 1 a.m.	40
1 a.m. – 6 a.m.	1

While we did record the number of comments and +1s every message received, these forms of interaction and propagation are not considered any further in this analysis for their ambiguous meaning in marketing contexts as detailed above.

Aside from these directly measured attributes of a message, we also computed every message's sentiment. German language messages (from the local German G+ pages) were automatically translated using the Google Translate API[3]. While automatic translation might not always work perfectly, wrong translations would only increase statistical variation and thus lead to conservative hypothesis test results. Therefore, effects that can be found using automatic translation are very likely to indeed occur in the sample population [41].

All computations (and indeed the harvesting for that matter) were done using R version 3.0.2 [42] and the **plusser** extension package [43]. Computation and visualization was aided by the R packages **ggplot2** [44] and **MASS** [45]. As detailed above, sentiment analysis extends into two distinct areas: polarity and emotions. The used software package offers to compute the positive—negative ratio, the

---

[3] C.f. https://developers.google.com/translate/.

ratio of the absolute log-likelihoods of the message expressing a positive or negative sentiment, to measure polarity of a message. A value of 1 indicates a neutral statement, while values smaller than 1 point towards a negative statement.

Measuring the involved emotions is a little bit more complicated. Here, six dimensions of frequently occurring emotions have been identified and for every message the log-likelihood of it reflecting these emotions is computed. Table 6 gives an overview of the distribution of the sentiment variables across the entire data set.

Table 6. Distribution parameters of message sentiments

Variable	Min	Mean	Max	SD
Polarity	0.04	16.13	111.39	22.04
Anger	1.47	1.53	7.34	0.59
Disgust	3.09	3.44	7.34	1.17
Fear	2.07	2.09	7.34	0.37
Joy	1.03	2.89	19.97	3.45
Sadness	1.73	2.12	7.34	1.43
Surprise	2.79	3.17	11.89	1.31

There is a considerable difference in the frequency and volume of messages being reshared between both countries, as is demonstrated by Fig. 1. In order to explain – at least partially – these differences, we use negative binomial models with a logistic link function.

The negative binomial model is an extension to the more familiar Poisson regression model. The latter frequently suffers from overdispersion, that is its sample variance exceeds its sample mean. It is possible to justify the appropriateness of a negative binomial model over a Poisson regression model, using a χ^2-test [45].

In order to model the reshare rate of a message (as opposed to raw counts of reshares), one has to take into account the potential exposure of a message. This is directly affected by the number of followers a G+ page has and the age of the message: the more followers to a page and the longer the message has been online, the more people are likely to have come across it. These measurement windows or base references are included as log offset terms in the models [45].

For developing the model, we combine model enhancement with backward selection based on AIC. We start out with a simple model using the classical covariates as described above and optimizing it using backward selection (M1). In a next step, we introduce *Country* (US being the reference category) as an interaction variable and employ backward selection again (M2). Finally, we include the variables from our sentiment analysis (M3). All models were tested for the appropriateness of a negative binomial specification using aforementioned χ^2-test.

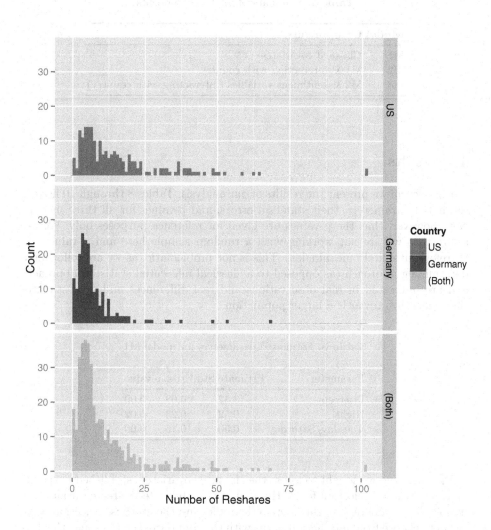

Fig. 1. Comparing the absolute number of reshares accross countries

In all models, the logarithm of message age is used as an offset to model exposure. For models that do not contain the variable *country*, the logarithm of the number of followers of a G+ page is included as an additional offset.

Table 7. Summary of model components

Model Components	
M1	Classical covariates
M2	M1 & interacting with country
M3	M2 & sentiment variables (interacting with country)

4 Results

In this section we present the results of our analysis. Tables 8 through 10 list the estimated parameters, their standard errors, and p-values for all three models. Note however, that the p-values are given for reference purposes only. We are aware that we are not working with a random sample here and p-values are therefore ultimately meaningless. This is not problematic, as we are following a data mining approach (as opposed to a classical inferential statistics approach) and are interested in real observations and real differences and not estimates that generalize towards a larger population.

Table 8. Estimated parameters for model M1

Parameter	Estimate	Std.Error	p.value
Intercept	-13.42	0.08	0.00
Night	0.52	0.23	0.02
Tuesday/Saturday	0.50	0.15	0.00

In a first step we sought to assess the classical covariates established in the literature. Stepwise selection found that only time- and date-related covariates are important parameters at this step. When allowing the stepwise selection algorithm to choose from interaction terms with the country of the G+ page, the day-time related term in the model becomes dependent on the country. The change in sign is a clear indication for the plausibility of the additional interaction. Model fit increased as well with $AIC_{M2} = 3004$ compared to $AIC_{M1} = 3113$.

The addition of the message sentiment variables improves the fit only marginally: $AIC_{M3} = 2994$. However, two sentiment variables' effects on the reshare rate differ across countries: Anger and Surprise.

Table 9. Estimated parameters for model M2

Parameter	Estimate	Std.Error	p.value
Intercept	-1.30	0.11	0.00
Evening/Night	1.02	0.25	0.00
Country	-1.26	0.15	0.00
Tuesday/Saturday	0.71	0.14	0.00
Evening/Night x Country	-0.55	0.31	0.08

Table 10. Estimated parameters for model M3

Parameter	Estimate	Std.Error	p.value
Intercept	-2.37	0.40	0.00
Evening/Night	1.07	0.24	0.00
Country	-0.01	0.46	0.98
Tuesday/Saturday	0.72	0.14	0.00
Disgust	0.11	0.06	0.06
Anger	0.44	0.21	0.04
Surprise	0.01	0.08	0.92
Evening/Night x Country	-0.70	0.31	0.02
Anger x Country	-0.44	0.24	0.07
Surprise x Country	-0.19	0.10	0.06

5   Discussion

These models allow for a number of insights. Foremost, we find in line with most literature on message propagation in social networks, that time of day has a highly significant effect on the number of rebroadcasts a message will receive. It is noteworthy, however, that the size of the effect changes with the country of the observation. Apparently, German G+ pages don't benefit from nighttime posts as much as Americans do. Table 11 presents the marginal effects of posting time across country.

Day of week also has a very significant influence here. Messages sent out on Tuesdays or Saturdays will receive increased attention. This effect is consistent across all models and does not depend on country.

Table 11. Marginal effect of a daytime post on expected reshare count across countries

Country	Daytime	Nighttime
Germany	7.42	10.73
US	27.15	79.24

Finally, there is another effect here that is dependent on culture: the role of emotion, more precisely of Anger and Surprise. There is a rather strong effect testifying to the German distaste of surprises. The expected count of reshares for a message increases with its notion of surprise in the US, while in Germany surprises slightly decrease these chances. A similar observation holds for Anger. Fig. 2 exhibits the sizes of these effects.

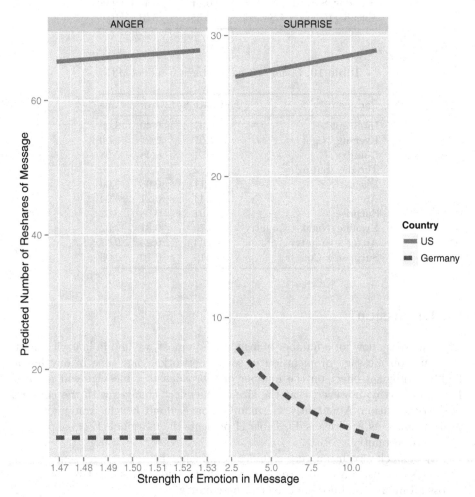

Fig. 2. Marginal effects of Anger and Surprise on expected reshare counts for US and German G+ pages

In light of the differences in culture observed by Hofstede, we found evidence for message traits that are consistent across diverse cultures differing in the individuality and uncertainty domains. This is chiefly the preference of certain weekdays and the perhaps obvious rejection of disgusting emotions embedded in advertising messages.

Other message and communication properties, time of day when the message was sent and angry or surprising sentiments, differ between the two cultures. Consumers in the US seem to prefer messages that are being sent out at nighttime, while their German counterparts are rather indifferent in that regard. This might be a sign for America's highly valued individualism and work ethics that would postpone past-time activities like checking Google+ pages to the nighttime.

German distaste of surprises and angry messages can as well be explained in terms of Hofstede's cultural dimensions: Uncertainty is something that is immanent to surprises and also anger poses certain risks. Germany's high level of uncertainty avoidance might explain the increased resharing of messages exhibiting neither anger nor surprise.

6 Conclusion

In light of these findings, it becomes clear that marketers must consider the culture of their audience even in an online arena. Alas, as Kitchen puts it:

> "Social media can be a proxy for consumer ethnography, the anthropological approach of understanding a culture by becoming part of it, because they provide virtual access to an often unguarded engagement with quite intimate aspects of consumer experience." [5, p. 36]

And who would not be cautious when treading with marketing intentions in intimate places? Cross-cultural communication might be hindered by a number of factors: ethnocentrism (i.e. ignoring other cultures), parochialism (focusing too much on local peculiarities) and stereotyping (because reality can be to complex to model) [46].

While it has become clear in the last few decades that ethnocentrism is a deadly sin for marketers and parochialism equally hinders effective campaigns, stereotyping is still all too familiar in marketing endeavors. After all, segmented target *groups* are defined by broad *averages* that marketers seek to please. And indeed, to avoid stereotyping, one needs not only to know the culture of a market, but to respect and adapt to individual consumers, wherever possible [12].

We believe that this adaption is easier to be had than previously thought: with the massive amounts of data available from social media platforms, marketers can seek to understand consumers more directly, away from cultural considerations, and just watch them forwarding and endorsing messages. It is this observation that empowers us to learn from consumers what they want.

References

1. Vargo, S.L., Lusch, R.F.: Evolving to a new dominant logic for marketing. Journal of Marketing 68(1), 1–17 (2004)
2. Merz, M.A., He, Y., Vargo, S.L.: The evolving brand logic: A service-dominant logic perspective. Journal of the Academy of Marketing Science 37(3), 328–344 (2009)

3. Chevalier, J.A., Mayzlin, D.: The effect of word of mouth on sales. Journal of Marketing Research 43(3), 345–354 (2006)
4. Rajagopal: Managing Social Media and Consumerism. Palgrave Macmillan (2013)
5. Kitchen, P.J.: The Dominant Influence of Marketing in the 21st Century: The Marketing Leviathan. Palgrave Macmillan (2013)
6. Kozinets, R.V., Hemetsberger, A., Schau, H.J.: The wisdom of consumer crowds: Collective innovation in the age of networked marketing. Journal of Macromarketing 28(4), 339–354 (2008)
7. Sánchez-Tabernero, A., Villanueva, J., Orihuela, J.L.: Social media networks as marketing tools for media companies. In: Handbook of Social Media Management, pp. 161–178. Springer (2013)
8. Alden, D.L., Steenkamp, J.B.E., Batra, R.: Brand positioning through advertising in Asia, North America, and Europe: The role of global consumer culture. The Journal of Marketing 63(1), 75–87 (1999)
9. Kotler, P., Pfoertsch, W.: B2B Brand Management. Springer, Berlin (2006)
10. Matthiesen, I., Phau, I.: The Hugo Boss connection: Achieving global brand consistency across countries. The Journal of Brand Management 12(5), 325–338 (2005)
11. Roth, M.S.: The effects of culture and socioeconomics on the performance of global brand image strategies. Journal of Marketing Research 32(2), 163–175 (1995)
12. Torelli, C.J.: Globalization, Culture, and Branding. Palgrave Macmillan (2013)
13. Raab, D.M.: New metrics for social media. Information Management 21(6), 24 (2011)
14. Ganahl, R.: The social media war: Is Google+ the David to Facebook's Goliath? In: Handbook of Social Media Management, pp. 513–531. Springer (2013)
15. Viégas, F., Wattenberg, M., Hebert, J., Borggaard, G., Cichowlas, A., Feinberg, J., Orwant, J., Wren, C.: Google+ ripples: A native visualization of information flow. In: Proceedings of the 22nd International Conference on World Wide Web, pp. 1389–1398. International World Wide Web Conferences Steering Committee (2013)
16. Kwak, H., Lee, C., Park, H., Moon, S.B.: What is Twitter, a social network or a news media? In: Proceedings of the 19th International Conference on World Wide Web, WWW 2010, pp. 591–600. ACM (2010)
17. Java, A., Song, X., Finin, T., Tseng, B.: Why we twitter: An analysis of a microblogging community. In: Zhang, H., Spiliopoulou, M., Mobasher, B., Giles, C.L., McCallum, A., Nasraoui, O., Srivastava, J., Yen, J. (eds.) WebKDD 2007. LNCS, vol. 5439, pp. 118–138. Springer, Heidelberg (2009)
18. Sago, B.: Factors influencing social media adoption and frequency of use: An examination of Facebook, Twitter, Pinterest and Google+. International Journal of Business & Commerce 3(1), 1–14 (2013)
19. Goyal, S.: Facebook, Twitter, Google+: Social networking. International Journal of Social Networking and Virtual Communities 1(1), 16–18 (2012)
20. McNair, B.: An Introduction to Political Communication. Routledge (2011)
21. Dennhardt, S.: User-Generated Content and its Impact on Branding. Springer (2012)
22. Rogers, E.M.: Diffusion of Innovations. Free Press (2010)
23. Shannon, C., Weaver, W.: The Mathematical Theory of Communication. University of Illinois Press (2002)
24. Vos, E., Varey, R.J.: Social networks and marketing happiness? The potential role of marketing in an electronic world. In: Varey, R.J., Pirson, M. (eds.) Humanistic Marketing, pp. 244–256 (2013)
25. Kempe, D., Kleinberg, J.M., Tardos, É.: Maximizing the spread of influence through a social network. In: Proceedings of the Ninth ACM SIGKDD International Conference on Knowledge Discovery and Data Mining, KDD 2003, pp. 137–146. ACM (2003)

26. Hochreiter, R., Waldhauser, C.: A stochastic simulation of the decision to retweet. In: Perny, P., Pirlot, M., Tsoukiàs, A. (eds.) ADT 2013. LNCS, vol. 8176, pp. 221–229. Springer, Heidelberg (2013)
27. Cha, M., Haddadi, H., Benevenuto, F., Gummadi, P.K.: Measuring user influence in Twitter: The million follower fallacy. In: Proceedings of the Fourth International Conference on Weblogs and Social Media, ICWSM 2010. The AAAI Press (2010)
28. Stieglitz, S., Dang-Xuan, L.: Political communication and influence through microblogging. In: 45th Hawaii International International Conference on Systems Science (HICSS-45 2012), pp. 3500–3509. IEEE Computer Society (2012)
29. Macskassy, S.A., Michelson, M.: Why do people retweet? Anti-Homophily wins the day! In: Proceedings of the Fifth International Conference on Weblogs and Social Media. The AAAI Press (2011)
30. Hofstede, G., Bond, M.H.: Hofstede's culture dimensions: An independent validation using Rokeach's value survey. Journal of Cross-Cultural Psychology 15(4), 417–433 (1984)
31. Taras, V., Kirkman, B.L., Steel, P.: Examining the impact of culture's consequences: A three-decade, multilevel, meta-analytic review of Hofstede's cultural value dimensions. Journal of Applied Psychology 95(3), 405–439 (2010)
32. Nakata, C.: Beyond Hofstede: Culture Frameworks for Global Marketing and Management. Palgrave Macmillan (2009)
33. Hofstede, G., Hofstede, G.J., Minkov, M.: Cultures and Organizations: Software for the Mind (2010)
34. Liu, B.: Sentiment Analysis and Opinion Mining. Morgan & Claypool Publishers (2012)
35. Hu, M., Liu, B.: Mining opinion features in customer reviews. In: Proceedings of Nineteeth National Conference on Artificial Intellgience (AAAI 2004). The AAAI Press (July 2004)
36. Bethard, S., Yu, H., Thornton, A., Hativassiloglou, V., Jurafsky, D.: Automatic extraction of opinion propositions and their holders. In: AAAI Spring Symposium on Exploring Attitude and Affect in Text 2004. The AAAI Press (2004)
37. Parrott, W.G.: Emotions in social psychology: Volume overview. In: Parrott, W.G. (ed.) Emotions in Social Psychology: Essential Readings, pp. 1–19. Psychology Press (2001)
38. Jurka, T.P.: Sentiment: Tools for Sentiment Analysis, R package version 0.2 (2012)
39. Guerzoni, M.: Product Variety in Automotive Industry. Springer (2014)
40. Schuman, J.H.: The Impact of Culture of Relationship Marketing in International Services. Springer (2009)
41. Brown, P.F., Della Pietra, V.J., Della Pietra, S.A., Mercer, R.L.: The mathematics of statistical machine translation. Computational Linguistics 19(2), 263–311 (1993)
42. R Core Team: R: A Language and Environment for Statistical Computing. R Foundation for Statistical Computing, Vienna, Austria (2013)
43. Waldhauser, C.: Plusser: A Google+ Interface for R. KDSS K Data Science Solutions, Vienna, Austria, R package version 0.3-5 (2014)
44. Wickham, H.: ggplot2: Elegant Graphics for Data Analysis. Springer, New York (2009)
45. Venables, W.N., Ripley, B.D.: Modern Applied Statistics with S. Springer, New York (2002)
46. Hurn, B.J., Tomalin, B.: Cross-Cultural Communication: Theory and Practice. Palgrave Macmillan (2013)

A Methodology for the Diagnostic of Aircraft Engine Based on Indicators Aggregation

Tsirizo Rabenoro[1,*], Jérôme Lacaille[1], Marie Cottrell[2], and Fabrice Rossi[2]

[1] Snecma, Groupe Safran,
77550 Moissy Cramayel, France
[2] SAMM (EA 4543), Université Paris 1,
90, rue de Tolbiac, 75634 Paris Cedex 13, France
{tsirizo.rabenoro,jerome.lacaille}@snecma.fr,
{marie.cottrell,fabrice.rossi}@univ-paris1.fr

Abstract. Aircraft engine manufacturers collect large amount of engine related data during flights. These data are used to detect anomalies in the engines in order to help companies optimize their maintenance costs. This article introduces and studies a generic methodology that allows one to build automatic early signs of anomaly detection in a way that is understandable by human operators who make the final maintenance decision. The main idea of the method is to generate a very large number of binary indicators based on parametric anomaly scores designed by experts, complemented by simple aggregations of those scores. The best indicators are selected via a classical forward scheme, leading to a much reduced number of indicators that are tuned to a data set. We illustrate the interest of the method on simulated data which contain realistic early signs of anomalies.

Keywords: Health Monitoring, Turbofan, Fusion, Anomaly Detection.

1 Introduction

Aircraft engines are generally made extremely reliable by their conception process and thus have low rate of operational events. For example, in 2013, the CFM56-7B engine, produced jointly by Snecma and GE aviation, has a rate of in flight shut down (IFSD) is 0.02 (per 1000 Engine Flight Hour) and a rate of aborted take-off (ATO) is 0.005 (per 1000 departures). This dispatch availability of nearly 100 % (99.962 % in 2013) is obtained also via regular maintenance operations but also via engine health monitoring (see also e.g. [15] for an external evaluation).

This monitoring is based, among other, on data transmitted by satellites[1] between aircraft and ground stations. Typical transmitted messages include engine

---

* This study is supported by a grant from Snecma, Safran Group, one of the world's leading manufacturers of aircraft and rocket engines, see http://www.snecma.com/ for details.
[1] Using the commercial standard Aircraft Communications Addressing and Reporting System (ACARS, see http://en.wikipedia.org/wiki/ACARS), for instance.

P. Perner (Ed.): ICDM 2014, LNAI 8557, pp. 144–158, 2014.

status overview as well as useful measurements collected as specific instants (e.g., during engine start). Flight after flight, measurements sent are analyzed in order to detect anomalies that are early signs of degradation. Potential anomalies can be automatically detected by algorithms designed by experts. If an anomaly is confirmed by a human operator, a maintenance recommendation is sent to the company operating the engine.

As a consequence, unscheduled inspections of the engine are sometimes required. These inspections are due to the abnormal measurements. Missing a detection of early signs of degradation can result in an IFSD, an ATO or a delay and cancellation (D&C). Despite the rarity of such events, companies need to avoid them to minimize unexpected expenses and customers' disturbance. Even in cases where an unscheduled inspection does not prevent the availability of the aircraft, it has an attached cost: it is therefore important to avoid as much as possible useless inspections.

We describe in this paper a general methodology to built complex automated decision support algorithms in a way that is comprehensible by human operators who take final decisions. The main idea of our approach is to leverage expert knowledge in order to build hundreds of simple binary indicators that are all signs of the possible existence of an early sign of anomaly in health monitoring data. The most discriminative indicators are selected by a standard forward feature selection algorithm. Then an automatic classifier is built on those features. While the classifier decision is taken using a complex decision rule, the interpretability of the features, their expert based nature and their limited number allows the human operator to at least partially understand how the decision is made. It is a requirement to have a trustworthy decision for the operator.

We will first describe the health monitoring context in Section 2. Then, we will introduce in more details the proposed methodology in Section 3. Section 4 will be dedicated to a simulation study that validates our approach.

2 Context

2.1 Flight Data

Engine health monitoring is based in part on flight data acquisition. Engines are equipped with multiple sensors which measure different physical quantities such as the high pressure core speed (N2), the Fuel Metering Valve (FMV), the Exhausted Gas Temperature (EGT), etc. (See Figure 1.) Those measures are monitored in real time during the flight. For instance the quantities mentioned before (N2, FMV, etc.) are analyzed, among others, during the engine starting sequence. This allows one to check the good health of the engine. If potential anomaly is detected, a diagnostic is made. Based on the diagnostic sent to a company operator, the airline may have to postpone the flight or cancel it, depending on the criticality of the fault and the estimated repair time.

The monitoring can also be done flight after flight to detect any change that can be flagged as early signs of degradations. Flight after flight, measurements are compressed in order to obtain an overview of engines status that consists

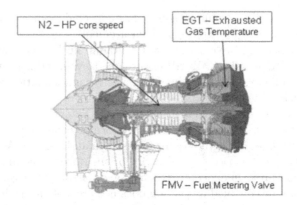

Fig. 1. Localization of some followed parameters on the Engine

in useful measurements at specific recurrent moments. These useful measurements are then preprocessed to obtain measurements independent from external environment. These preprocessed data are analyzed by algorithms and human operators. The methodology introduced in this article is mostly designed for this kind of monitoring.

2.2 Detecting Faults and Abnormal Behaviors

Traditional engine health monitoring is strongly based on expert knowledge and field experience (see e.g. [14] for a survey and [5] for a concrete example). Faults and early signs of faults are identified from suitable measurements associated to adapted computational transformation of the data. For instance, the different measurements (temperatures, vibration, etc.) are influenced by the flight parameters (e.g. throttle position) and conditions (outside temperature, etc.). Variations in the measured values can therefore result from variations in the parameters and conditions rather than being due to abnormal behavior. Thus a typical computational transformation consists in preprocessing the measurements in order to remove dependency to the flight context [10].

In practice, the choice of measurements and computational transformations is generally done based on expert knowledge. For instance in [12], a software is designed to record expert decision about a time interval on which to monitor the evolution of such a measurement (or a time instant when such a measurement should be recorded). Based on the recorded examples, the software calibrates a pattern recognition model that can automatically reproduce the time segmentation done by the expert. Once the indicators have been computed, the normal behavior of the indicators can be learned. The residuals between predictions and actual indicators can be statistically modeled, e.g. as a Gaussian vector. A score measurement is obtained from the likelihood of this distribution. The normalized vector is a failure score signature that may be described easily by experts to identify the fault origin, in particular because the original indicators have some meaning for them. See [4], [5] and [9] for other examples.

However experts are generally specialized on a particular subsystem, thus each algorithm focuses mainly on a specific subsystem despite the need of a diagnostic of the whole system.

2.3   Data and Detection Fusion

The global diagnostic is currently done by the operator who collects all available results of diagnostic applications. The task of taking a decision based on all incoming information originating from different subsystems is difficult. A first difficulty comes from dependencies between subsystems which means that for instance in some situations, a global early sign of failure could be detected by discovering discrepancies between seemingly perfectly normal subsystems. In addition, subsystem algorithms can provide conflicting results or a decision with a very low confidence level. Furthermore, extreme reliabilities of engines lead to an exacerbated trade off between false alarm levels and detection levels, leading in general to a rather high level of false alarms, at least at the operator level. Finally, the role of the operator is not only to identify a possible early sign of failure, but also to issue recommendations on the type of preventive maintenance needed. In other words, the operator needs to identify the possible cause of the potential failure.

2.4   Objectives

The long term goal of engine health monitoring is to reach automated accurate, trustworthy and precise maintenance decisions during optimally scheduled shop visit, but also to drastically reduce operational events such as IFSD and ATO. However, partly because of the current industrial standard, pure black box modeling is unacceptable. Indeed, operators are currently trained to understand expertly designed indicators and to take complex integrated decisions on their own. In order for a new methodology to be accepted by operators, it has at least to be of a gray box nature, that is to be (partially) explainable via logical and/or probabilistic reasoning. Then, our objective is to design a monitoring methodology that helps the human operator by proposing integrated decisions based on expertly designed indicators with a "proof of decision".

3   Architecture of the Decision Process

3.1   Health Monitoring Data

In order to present the proposed methodology, we first describe the data obtained via health monitoring and the associated decision problem.

We focus here on ground based long term engine health monitoring. Each flight produces dozens of timestamped flight events and data. Concatenating those data produces a multivariate temporal description of an engine whose dimensions are heterogeneous. In addition, sampling rates of individual dimensions might be different, depending on the sensors, the number of critical time points recorded in a flight for said sensor, etc.

Based on expert knowledge, this complex set of time series is turned into a very high dimensional indicator vector. The main idea, outlined in the previous section, is that experts generally know what is the expected behavior of a subsystem of the engine during each phase of the flight. Then the dissimilarity between the expected behavior and the observed one can be quantified leading to one (or several) anomaly scores. Such scores are in turn transformed into binary indicators where 1 means an anomaly is detected and 0 means no anomaly detected.

This transformation has two major advantages: it homogenizes the data and it introduces simple but informative features (each indicator is associated to a precise interpretation related to expert knowledge). It leads also to a loss of information as the raw data are in general non recoverable from the indicators. This is considered here a minor inconvenience as long as the indicators capture all possible failure modes. This will be partially guaranteed by including numerous variants of each indicator (as explained below). On a longer term, our approach has to be coupled with field experience feedback and expert validation of its coverage.

After the expert guided transformation, the monitoring problem becomes a rather standard classification problem: based on the binary indicators, the decision algorithm has to decide whether there is an anomaly in the engine and if, this is the case, to identify the type of the anomaly (for instance by identifying the subsystem responsible for the potential problem).

We describe now in more details the construction of the binary indicators.

3.2   Some Types of Anomalies

Some typical univariate early signs of anomalies are shown on Figures 2, 3 and 4 which display the evolution through time of a numerical value extracted from real world data. One can identify, with some practice, a variance shift on Figure 2, a mean shift on Figure 3 and a trend modification (change of slope) on Figure 4. In the three cases, the change instant is roughly at the center of the time window.

The main assumption used by experts in typical situations is that, when external sources of change have been accounted for, the residual signal should be stationary in a statistical sense. That is, observations

$$\mathcal{Y}_n = (Y_1(\theta_1), ..., Y_n(\theta_n))$$

are assumed to be generated identically and independently from a fixed parametric law, with a constant set of parameters (that is, all the θ_i are identical). Then, detecting an anomaly amounts to detecting a change in the time series (as illustrated by the three Figures above). This can be done via numerous well known statistical tests [1]. In the multivariate cases, similar shifts in the signal can be associated to anomalies. More complex scenarios, involving for instance time delays, can also be designed by experts.

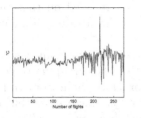
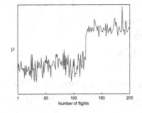
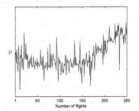

Fig. 2. Variance shift   **Fig. 3.** Mean shift   **Fig. 4.** Trend modification

3.3   From Anomaly Types to Indicators

While experts can generally describe explicitly what type of change they are expecting for some specific early signs of anomaly, they can seldom provide detailed parameter settings for statistical tests (or even for the aggregation technique that could lead to a statistical test after complex calculations). To maximize coverage it seems natural to include numerous indicators based on variations of the anomaly detectors compatible with expert knowledge.

Let us consider for illustration purpose that the expert recommends to look for shifts in mean of a certain quantity as early signs of a specific anomaly. If the expert believes the quantity to be normally distributed with a fixed variance, then a natural test would be Student's t-test. If the expert has no strong priors on the distribution, a natural test would be the Mann–Whitney U test. Both can be included to maximize coverage.

Then, in both cases, one has to assess the scale of the shift. Indeed those tests work by comparing summary statistics of two populations, before and after a possible change point. To define the populations, the expert has to specify the length of the time windows to consider before and after the possible change point: this is the expected scale at which the shift will appear. In most cases, the experts can only give a rough idea of the scale. Again, maximizing the coverage leads to the inclusion of several scales compatible with the experts' recommendations.

Given the choice of the test, of its scale and of a change point, one can construct a statistic. A possible choice for the indicator could be this value or the associated p-value. However, we choose to use simpler indicators to ease their interpretation. Indeed, the raw value of a statistic is generally difficult to interpret. A p-value is easier to understand because of the uniform scale, but can still lead to misinterpretation by operators with insufficient statistical training. We therefore choose to use binary indicators for which the value 1 corresponds to a rejection of the null hypothesis of the underlying test to a given level (the null hypothesis is here the case with no mean shift).

Finally, as pointed out before, aircraft engines are extremely reliable, a fact that increases the difficulty in balancing sensibility and specificity of anomaly detectors. In order to alleviate this difficulty, we build high level indicators from low level tests. For instance, if we monitor the evolution of a quantity on a long period compared to the expected time scale of anomalies, we can compare the number of times the null hypothesis of a test has been rejected on the long period with the number of times it was not rejected, and turn this into a binary indicator with a majority rule.

To summarize, we construct parametric anomaly scores from expert knowledge, together with acceptable parameter ranges. By exploring those ranges, we generate numerous (possible hundreds of) binary indicators. Each indicator can be linked to an expertly designed score with a specific set of parameters and thus is supposedly easy to interpret by operators. Notice that while we as focused in this presentation on temporal data, this framework can be applied to any data source.

3.4   Decision

The final decision step consists in classifying these high dimensional binary vectors into at least two classes, i.e., the presence or absence of an anomaly. A classification into more classes is highly desirable if possible, for instance to further discriminate between seriousness of anomalies and/or sources (in terms of subsystems of the engine). In this paper however, we will restrict ourselves to a binary classification case (with or without anomaly).

As explained before, we aim in the long term at gray box modeling, so while numerous classification algorithms are available see e.g. [8], we shall focus on interpretable ones. In this paper, we choose to use Random Forests [2] as they are very adapted to binary indicators and to high dimensional data. They are also known to be robust and to provide state-of-the-art classification performances at a very small computational cost. While they are not as interpretable as their ancestors CART [3], they provide at least variable importance measures that can be used to identify the most important indicators.

Finally, while including hundreds of indicators is important to give a broad coverage of the parameter spaces of the expert scores and thus to maximize the probability of detecting anomalies, it seems obvious that some redundancy will appear. Therefore, we have chosen to apply a feature selection technique [6] to this problem. The reduction of number of features will ease the interpretation by limiting the quantity of information transmitted to the operators in case of a detection by the classifier. Among the possible solutions, we choose to use the Mutual information based technique Minimum Redundancy Maximum Relevance (mRMR, [11]) which was reported to give excellent results on high dimensional data.

4   A Simulation Study

4.1   Introduction

It is difficult to find real data with early signs of degradations, because their are scarce and moreover the scheduled maintenance operations tend to remove these early signs. Experts could study in detail recorded data to find early signs of anomalies whose origins were fixed during maintenance but it is close to looking for a needle in a haystack, especially considering the huge amount of data to analyze. We will therefore rely in this paper on simulated data. Our goal is to validate the interest of the proposed methodology in order to justify investing in the production of carefully labelled real world data.

In this section we begin by the description of the simulated data used for the evaluation of the methodology, and then we will present the performance obtained on this data.

4.2   Simulated Data

The simulated data are generated according to the univariate shift models described in Section 3.2. We generate two data sets a simple one A and a more complex one B.

In the first case A, it is assumed that expert based normalisation has been performed. Therefore when no shift in the data distribution occurs, we observe a stationary random noise modeled by the standard Gaussian distribution. Using notations of Section 3.2 the Y_i are independent and identically distributed according to $\mathcal{N}(\mu = 0, \sigma^2 = 1)$. Signals in set A have a length chosen uniformly at random between 100 and 200 observations (each signal has a specific length).

Anomalies are modelled after the three examples given in Figures 2, 3 and 4. We implement therefore three types of shift:

1. a variance shift: in this case, observations are distributed according to $\mathcal{N}(\mu = 0, \sigma^2)$ with $\sigma^2 = 1$ before the change point and σ chosen uniformly at random in $[1.01, 5]$ after the change point (see Figure 5);
2. a mean shift: in this case, observations are distributed according to $\mathcal{N}(\mu, \sigma^2 = 1)$ with $\mu = 0$ before the change point and μ chosen uniformly at random in $[1.01, 5]$ after the change point (see Figure 6);
3. a slope shift: in this case , observations are distributed according to $\mathcal{N}(\mu, \sigma^2 = 1)$ with $\mu = 0$ before the change point and μ increasing linearly from 0 from the change point with a slope chosen uniformly at random in $[0.02, 3]$ (see Figure 7).

Assume that the signal contains n observations, then the change point is chosen uniformly at random between the $\frac{2n}{10}$-th observation and the $\frac{8n}{10}$-th observation.

According to this procedure, we generate a balanced data set with 6000 observations corresponding to 3000 observations with no anomaly, and 1000 observations for each of the three types of anomalies.

In the second data set, B, a slow deterministic variation is added to randomly chosen signals with no anomaly: this is a way to simulate a suboptimal normalisation (see Figure 8 for an example). The slow variation is implemented by adding to the base noise a sinus with a period of $\frac{2}{3}$ of the signal length and amplitude 1.

Signals in set B are shorter, to make the detection more difficult: they are chosen uniformly at random between 100 and 150 observations. In addition, the noise is modeled by a χ^2 distribution with 4 degrees of freedom. Signals with an anomaly are generated using the same rationale as for set A. In this case however, the mean

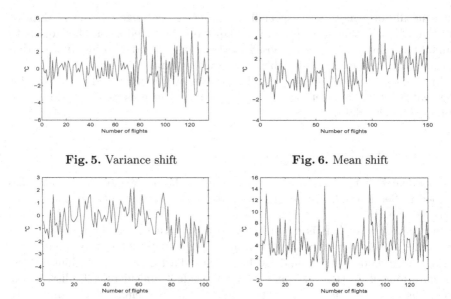

Fig. 5. Variance shift   **Fig. 6.** Mean shift

Fig. 7. Trend modification   **Fig. 8.** Data without anomaly but with suboptimal normalisation represented by a slow varying deterministic component

shift is simply implemented by adding a constant to the signal after the change point. The "variance" shift is in fact a change in the number of degrees of freedom of the χ^2 distribution: after the change point, the number of degrees is chosen randomly (uniformly) between 8 and 16. The change point is chosen as in set A.

According to this procedure, we generate a balanced data set with 6000 observations corresponding to 3000 observations with no anomaly, and 1000 observations for each of the three types of anomalies. Among the 3000 anomaly free signals, 1200 are corrupted by a slow variation.

4.3   Indicators

As explained in Section 3.3, binary indicators are constructed from expert knowledge by varying parameters, including scale and position parameters. In the present context, we use sliding windows: for each position of the window, a classical statistical test is conducted to decide whether a shift in the signal occurs at the center of the window.

The "expert" designed tests are here:

1. the Mann–Whitney–Wilcoxon U test (non parametric test for shift in mean);
2. the two sample Kolmogorov-Smirnov test (non parametric test for differences in distributions);
3. the F-test for equality of variance (parametric test based on a Gaussian hypothesis).

The direct parameters of those tests are the size of the window which defines the two samples (30, 50, and $\min(n - 2, 100)$ where n is the signal length) and the level of significance of the test (0.005, 0.1 and 0.5). Notice that those tests do not include a slope shift detection.

Then, more complex binary indicators are generated, as explained in Section 3.3. In a way, this corresponds to build very simple binary classifiers. We use the following ones:

1. for each underlying test, the derived binary indicator takes the value one if on a fraction β of m windows, the test detects a change. Parameters are the test itself with its parameters, the value of β (we considered 0.1, 0.3 and 0.5) and the number of observations in common between two consecutive windows (the length of the window minus 1, 5 or 10);
2. for each underlying test, the derived binary indicator takes the value one if on a fraction β of m consecutive windows, the test detects a change (same parameters);
3. for each underlying test, the derived binary indicator takes the value one if there are 5 consecutive windows such that the test detects a change on at least k of these 5 consecutive windows (similar parameters where β is replaced by k).

In addition, based on expert recommendation, we apply all those indicators both to the original signal and to a smoothed signal (using a simple moving average over 5 measurements).

We use more than 50 different configurations for each indicator, leading to a total number of 810 binary indicators (it should be noted that only a subset of all possible configurations is included into this indicator vector).

4.4   Reference Performances

In this paper, we focus on the simple case of learning to discriminate between a stationary signal and a signal with a shift. We report therefore the classification rate (classification accuracy).

For both sets A et B, the learning sample is composed of 1000 signals keeping the balance between the three classes of shifts. The evaluation is done on the remaining 5000 signals that have been divided in 10 groups of 500 time series each. The rationale of this data splitting in the evaluation phase is to estimate both the prediction quality of the model but also the variability in this rate as an indication of the trust we can put on the results. We also use and report the out-of-bag (OOB) estimate of the performances provided by the Random Forest (this is a byproduct of the bootstrap procedure used to construct the forest, see [2]).

When all the 810 indicators are used, the classification performances are very high on set A and acceptable on set B (see Table 1). The similarity between the OOB estimate of the performances and the actual performances confirms that the OOB performances can be trusted as a reliable estimator of the actual performances. Data set B shows a strong over fitting of the Random Forest, whereas data set A exhibits a mild one.

Table 1. Classification accuracy of the Random Forest using the 810 binary indicators. For the test set, we report the average classification accuracy and its standard deviation between parenthesis.

Data set	Training set accuracy	OOB accuracy	Test set average accuracy
A	1	0.953	0.957 (0.0089)
B	1	0.828	0.801 (0.032)

4.5   Feature Selection

On the data set A, the performances are very satisfactory, but the model is close to a black box in the sense that it uses all the 810 indicators. Random Forests are generally difficult to interpret, but a reduction in the number of indicators would allow an operator to study the individual decisions performed by those indicators in order to have a rough idea on how the global decision could have been made. On the data B set, the strong over fitting is another argument for reducing the number of features.

Using the mRMR we ranked the 810 indicators according to a mutual information based estimation of their predictive performances. We use then a forward approach to evaluate how many indicators are needed to achieve acceptable predictive performances. Notice that in the forward approach, indicators are added in the order given by mRMR and then never removed. As mRMR takes into account redundancy between the indicators, this should not be a major issue. Then for each number of indicators, we learn a Random Forest on the learning set and evaluate it.

Figure 9 shows the results for data set A. Accuracies are quite high with a rather low number of indicators, with a constant increase in performances on the learning set (as expected) and a stabilisation of the real performances (as evaluated by the test set and the OOB estimation) around roughly 40 indicators.

Figure 10 shows the results for data set B. Excepted the lower performances and the stronger over fitting, the general behavior is similar to the case of data set A. Those lower performances were expected because data set B has been designed to be more difficult to analyze, in part because of the inadequacy between the actual data model (χ^2 distribution) and the "expert" based low level tests (in particular the F test which assumes Gaussian distributions).

In both cases, the feature selection procedure shows that only a small subset of the original 810 indicators is needed to achieve the best performances reachable on the data sets. This is very satisfactory as this allows to present to operators a manageable number of binary decisions together with the aggregated one provided by the random forest.

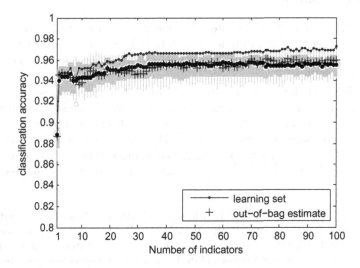

Fig. 9. Data set A : classification accuracy on learning set (circle) as a function of the number of indicators. A boxplot gives the classification accuracies on the test subsets, summarized by its median (black dot inside a white circle). The estimation of those accuracies by the out-of-bag (OOB) bootstrap estimate is shown by the crosses.

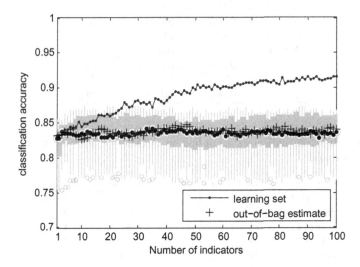

Fig. 10. Data set B : classification accuracy on learning set (circle) as a function of the number of indicators. A boxplot gives the classification accuracies on the test subsets, summarized by its median (black dot inside a white circle). The estimation of those accuracies by the out-of-bag (OOB) bootstrap estimate is shown by the crosses.

4.6   Selected Indicators

In order to illustrate further the interest of the proposed methodology, we show in Table 2 the best ten indicators for data set A. Those indicators lead to quite good performances with an average test set classification accuracy of 0.944 (OOB estimation is 0.938).

Table 3 shows the best ten indicators for data set B. Again, this corresponds to quite good performances with an average test set classification accuracy of 0.831 (OOB estimation is 0.826). As expected, the F test and indicators based on it are less interesting for this data set as the noise is no more Gaussian.

In both cases, we see that the feature selection method is able to make a complex selection in a very large set of binary indicators. This induces indirectly an

Table 2. The best ten indicators according to mRMR for data set A. Confu(k,n) corresponds to a positive Mann–Whitney–Wilcoxon U test on k windows out of n consecutive ones. Conff(k,n) is the same thing for the F-test. Ratef(α) corresponds to a positive F-test on $\alpha \times m$ windows out of m. Lseqf(α) corresponds to a positive F-test on $\alpha \times m$ consecutive windows out of m. Lsequ(α) is the same for a U test. Here, none of the indicators are based on a smoothed version of the signal.

type of indicator	level	window length	window step
F test	0.005	100	5
confu(2,3)	0.005	50	5
ratef(0.1)	0.005	50	5
KS test	0.005	100	1
conff(3,5)	0.005	100	5
KS test	0.1	100	5
F test	0.005	100	1
KS test	0.005	100	10
lseqf(0.1)	0.1	50	1
F test	0.005	50	10

Table 3. The best ten indicators according to mRMR for data set B. Please refer Table 2 for notations.

type of indicator	level	window length	smoothed	window step
KS test	0.005	100	no	5
lseqf(0.1)	0.1	30	yes	1
confu(4,5)	0.005	30	no	1
U test	0.1	100	no	5
confu(4,5)	0.005	100	no	5
confu(2,3)	0.005	100	no	1
lsequ(0.3)	0.1	50	no	1
F test	0.005	100	yes	10
confu(2,3)	0.005	30	no	5
KS Test	0.005	100	no	10

automatic tuning of the parameters of the low level tests and of simple aggregation classifiers. Because of their simplicity and their binary outputs, indicators
are easy to understand by an operator.

5 Conclusion and Perspectives

In this paper, we have introduced a diagnostic methodology for engine health
monitoring that leverage expert knowledge and automatic classification. The
main idea is to build from expert knowledge parametric anomaly scores associated to range of plausible parameters. From those scores, hundreds of binary
indicators are generated in a way that covers the parameter space as well as introduces simple aggregation based classifiers. This turns the diagnostic problem
into a classification problem with a very high number of binary features. Using
a feature selection technique, one can reduce the number of useful indicators to
a humanly manageable number. This allows a human operator to understand
at least partially how a decision is reached by an automatic classifier. This is
favored by the choice of the indicators which are based on expert knowledge and
on very simple decision rules. A very interesting byproduct of the methodology
is that is can work on very different original data as long as expert decision
can be modelled by a set of parametric anomaly scores. This was illustrated by
working on signals of different lengths.

Using simulated data, we have shown that the methodology is sound: it reaches
good predictive performances even with a limited number of indicators (e.g., 10).
In addition, the selection process behaves as expected, for instance by discarding
statistical tests that are based on hypothesis not fulfilled by the data. However,
we limited ourselves to univariate data and to a binary classification setting
(i.e., abnormal versus normal signal). We need to show that the obtained results
can be extended to multivariate data and to complex classification settings (as
identifying the cause of a possible anomaly is extremely important in practice).

It should also be noted that we relied on Random Forests which are not as
easy to interpret as other classifiers (such as CART). In our future work, we will
compare Random Forest to simpler classifiers. As we are using binary indicators,
some form of majority voting is probably the simplest possible rule but using
such as rule implies to choose very carefully the indicators [13].

Finally, it is important to notice that the classification accuracy is not the
best way of evaluating the performances of a classifier in the health monitoring
context. Firstly, health monitoring intrinsically involves a strong class imbalance
[7]. Secondly, health monitoring is a cost sensitive area because of the strong
impact on airline profit of an unscheduled maintenance. It is therefore important
to take into account specific asymmetric misclassification cost to get a proper
performance evaluation.

Acknowledgment. This study is supported by a grant from Snecma, Safran
Group, one of the world's leading manufacturers of aircraft and rocket engines,
see http://www.snecma.com/ for details.

References

1. Basseville, M., Nikiforov, I.V.: Detection of abrupt changes: theory and applications. Journal of the Royal Statistical Society-Series A Statistics in Society 158(1), 185 (1995)
2. Breiman, L.: Random forests. Machine Learning 45(1), 5–32 (2001)
3. Breiman, L., Friedman, J.H., Olshen, R.A., Stone, C.J.: Classification and regression trees. Wadsworth & Brooks, Monterey (1984)
4. Côme, E., Cottrell, M., Verleysen, M., Lacaille, J.: Aircraft engine health monitoring using self-organizing maps. In: Perner, P. (ed.) ICDM 2010. LNCS, vol. 6171, pp. 405–417. Springer, Heidelberg (2010)
5. Flandrois, X., Lacaille, J., Masse, J.R., Ausloos, A.: Expertise transfer and automatic failure classification for the engine start capability system. AIAA Infotech, Seattle (2009)
6. Guyon, I., Elisseeff, A.: An introduction to variable and feature selection. The Journal of Machine Learning Research 3, 1157–1182 (2003)
7. Japkowicz, N., Stephen, S.: The class imbalance problem: A systematic study. Intelligent Data Analysis 6(5), 429–449 (2002)
8. Kotsiantis, S.B., Zaharakis, I., Pintelas, P.: Supervised machine learning: A review of classification techniques (2007)
9. Lacaille, J.: A maturation environment to develop and manage health monitoring algorithms. PHM, San Diego (2009)
10. Lacaille, J.: Standardized failure signature for a turbofan engine. In: 2009 IEEE Aerospace Conference, pp. 1–8. IEEE (2009)
11. Peng, H., Long, F., Ding, C.: Feature selection based on mutual information criteria of max-dependency, max-relevance, and min-redundancy. IEEE Transactions on Pattern Analysis and Machine Intelligence 27(8), 1226–1238 (2005)
12. Rabenoro, T., Lacaille, J.: Instants extraction for aircraft engine monitoring. AIAA Infotech@Aerospace (2013)
13. Ruta, D., Gabrys, B.: Classifier selection for majority voting. Information Fusion 6(1), 63–81 (2005)
14. Tumer, I.Y., Bajwa, A.: A survey of aircraft engine health monitoring systems. In: Proc. 35th Joint Propulsion Conf. (1999)
15. Vasov, L., Stojiljković, B.: Reliability levels estimation of JT8D-9 and CFM56-3 turbojet engines. FME Transactions 35(1), 41–45 (2007)

Visual Trajectory Pattern Mining: An Exploratory Study in Baggage Handling Systems

Ayman Al-Serafi[1] and Ahmed Elragal[2]

[1] Teradata Corporation
ayman.al-serafi@teradata.com
[2] Department of Business Informatics & Operations,
German University in Cairo (GUC), Cairo, Egypt
ahmed.elragal@guc.edu.eg

Abstract. There is currently a huge amount of data being collected about movements of objects. Such data is called spatiotemporal data and paths left by moving-objects are called trajectories. Recently, researchers have been targeting those trajectories for extracting interesting and useful knowledge by means of pattern analysis and data mining. But, it is difficult to analyse huge datasets of trajectories without summarizing them and visualizing them for the knowledge seeker and for the decision makers. Therefore, this research paper focuses on utilizing visual techniques and data mining analysis of trajectory patterns in order to help extract patterns and knowledge in an interactive approach. The research study proposes a research framework which integrates multiple data analysis and visualization techniques in a coherent architecture in support of interactive trajectory pattern visualization for the decision makers. An application case-study of the techniques is conducted on an airport's baggage movement data within the Baggage Handling System (BHS). The results indicate the feasibility of the approach and its methods in visually analysing trajectory patterns in an interactive approach which can support the decision maker.

Keywords: Trajectory Pattern Mining, Frequent Pattern Mining, Visual Analytics, Business Intelligence, Baggage Handling System Data.

1   Introduction

Currently there is an evolving issue concerning analysing movements of objects which involves overflow of data and limited capability to extract knowledge from this data. There is also high availability of time-stamped data commonly collected with devices that track moving-objects, which is called spatiotemporal data. Traces of movement left by moving objects and collected using spatiotemporal data gathering are called trajectories. A moving-object is any object which is continuously moving causing its locational position to change.

One of the main problems for businesses today, however, is that such raw trajectory data leads to little knowledge for decision makers if it is given to the

P. Perner (Ed.): ICDM 2014, LNAI 8557, pp. 159–173, 2014.
© Springer International Publishing Switzerland 2014

business decision makers without analysing them and extracting useful knowledge [4]. As a result, data mining techniques are commonly utilized in geospatial and trajectory analysis. And, as a consequence, it is also of relevance to be able to visualize trajectories and to extract interesting patterns from them in different applications.

One of the applications can be airports. Currently, there is a continuous requirement for more timely services and more efficient processes in airports which requires implementing information technologies [19]. There is also a great trend of transfer passengers which pass-by an airport to catch another flight. Such transfer passengers create an increasing demand for more efficient and swifter transfer passengers handling processes [19]. Metrics and business intelligence on baggage handling system (BHS) data is of increasing interest nowadays in order to meet higher levels of passenger satisfaction and to meet those challenging requirements [21]. Baggage data analysis was also recommended for future research in [23]. Therefore, this research paper explores the adoption of visual analytics in trajectory pattern analysis for baggage handling data, i.e. baggage movement tracking data. A qualitative case study of a major airport's data is used to apply a developed prototype to test the techniques.

BHS plays a critical role in airport operations, as it is a significant technology for processing passengers and is one of the most complex automated operational systems in an airport. The aim of the BHS system is to keep the baggage synchronized with the passengers and their flights and not to lag behind or surpass the movements of passengers. One of the main goals of the BHS which will be covered in this research paper is to move baggage from check-in counters to their allocated flight locations for departing passengers. In the case study of this research paper, the BHS transports baggage from sources (like airport check-in counters) to output locations (like the make-up chutes for loading baggage into containers for each flight). The BHS system does extra tasks as well like security baggage screening and scanning-in baggage items. This makes it important for BHS personnel to set goals and monitor key performance indicators (KPIs), which will be utilized in this research paper. For a detailed overview of the components and functions of the BHS refer to [11].

This paper is organized as follows: section 2 discusses the related work to visual analytics of trajectories and the BHS applications, section 3 introduces the main research framework implementing the visual analytics on trajectories, section 4 applies the research framework on a case study airport, and finally section 5 concludes the research paper.

2 Related Work: Visual Analytics of Trajectories

This research paper adopts data mining techniques in order to extract useful knowledge from baggage trajectories. A trajectory is a journey by a moving-object between two places (which can be considered as a starting and ending locations) [22]. A trajectory can also be formally defined for a moving object T as a series of spatiotemporal points having location spatial coordinates (X and Y) and time (t) which states the positions in space and time of recording the movement in space. This can be seen in figure 1 adapted from [2, p. 125] for the application of BHS at airports. For a typical automated BHS, the moving-object is a bag, the trajectory start location is the security level 1feedline and the end location can be any location from a lateral chute or a manual inspection station. This movement is tracked by barcode readers reading the tags on the bags. The BHS computer servers store data about the location and movements of baggage moving-objects. The data available from the BHS log-files include baggage tracking within screening feedlines and conveyors, and the sortation conveyors and laterals.

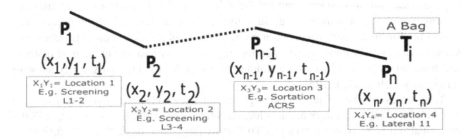

Fig. 1. A Baggage Trajectory inside the BHS

To analyse the baggage trajectories, frequent pattern mining (FPM) is used. FPM is a data mining strategy which includes techniques like association rule mining and sequence pattern mining [9]. Frequent patterns can be defined as: "Frequent patterns are itemsets, subsequences, or substructures that appear in a data set with frequency no less than a user-specified threshold" [9, p. 56]. A frequent itemset is a set of items associated by having high frequency together in a transactional database. Sequential pattern mining is a technique for finding frequently occurring ordered sub-sequences of items or events [9]. Association rule mining (ARM) aims to find associations and correlations between items in a transactional database [26]. The goal of ARM is to find strong rules with minimum support and minimum confidence [1]. Support measures the strength (statistical importance) of the association rule and the confidence measures the certainty of the association rule [1].

Previous research in applying FPM onto trajectories can be seen in research like [18] for people movement patterns, [25] for car-traffic data, and [7, 8, 26] for general spatiotemporal patterns. On the other hand, ARM can also be applied on trajectories to find relationships between patterns where the existence of a specific pattern leads to the existence of another pattern. For example, a trajectory consisting of eastward movement from the city centre at 10 P.M. results into westward movement towards the city centre at 1 A.M [7].

Visualizations are commonly used for exploratory data analysis and mining [14]. This is applied using visual analytics which is about the integration of the human expertise into the knowledge discovery and data exploration process [20]. Nowadays, interactive visualizations are gaining significant importance. Interactive visualization is the combination of human effort with computational effort to support the knowledge discovery process [5]. It commonly involves filters and aggregations which the users specify to reduce the visualizations to a subset which can be interpreted by human cognition [5, 14]. Interactive visualization requires high-level graphical user interfaces presenting data mining results and allowing for interaction with the end-user in terms of specifying instructions and commands that guide the pattern analysis process [14]. Filters based on specific parameters are a common way of enhancing visualizations of large spatiotemporal datasets.

Spatiotemporal visualization can be commonly used to track moving-objects and events related to the movements or stops of the moving-objects. Map visualizations are essential for any spatiotemporal visualization. Overlaying extra details and information about movements (e.g. direction of movement, speed of the object, etc.) is also used to make the visualization more meaningful [5]. Maps and 3D cubes also known as space-time cubes are then commonly used in visualizing movement patterns and trajectories [4, 7, 12]. Other examples of research utilizing visual analytics with trajectories include [5, 17, 25].

Trajectories can be visualized on a geospatial map where the trajectory paths are visualized as continuous lines and key locations (also called points of interests, or POIs) are visualized as reference marks [3]. To find frequent locations a density map can be used. A density map shows frequently visited locations as darker cells [5, p. 132]. Extra statistics (e.g. frequency rate and congestion rates) can be overlaid over the map to show further details about each location. Beside maps (also commonly called cartographic visualizations), it is also important to show aggregate data about movements in non-cartographic displays such as bar-charts and graphs [5, 6, & 15 p. 1144].

When FPM is applied to trajectories, it is called Trajectory Pattern Mining. For trajectory pattern mining, visual analytics can include interactive selection of the support minimum and maximum thresholds to use when looking for frequent trajectory patterns [3, 8]. Interactive selection of locations which the user is interested to see connectional patterns between is also common in moving-object pattern analysis [3]. Another type of interactive trajectory visualization makes the user select two locations and the visualization should show the frequent paths (patterns) available connecting those two locations together [12]. The timeframe within which a trajectory took place can be used as a filter by the user to see only relevant trajectory plots [8]. The temporal aspect is usually selected using a slider which helps the user scroll through time instances of a spatial map [12].

There are a few groups of research studies the researchers are aware of which focused around information systems (IS) in BHS used in airports of the aviation industry. For example, [11] is part of the BHS literature group related to design and routing optimization. Their approach uses artificial intelligence techniques.

One of the few research studies found examining the use of business intelligence in BHS was [10]. Heinz & Pitfield [10] introduced the concept of using KPIs to monitor baggage handling performance. This can be seen for example when they profile transfer baggage performance as a ratio between numbers of mishandled bags against total numbers of passengers. However, [10] didn't discuss details about data mining and visual analytics of baggage movement data integrated into their framework, which will be researched in this paper.

3   A Framework for Trajectory Visual Analysis

Currently, the main problem is that trajectory mining techniques reviewed lack sufficient interactive support for decision making. Therefore, handling expert interaction within data mining is important for yielding more relevant results by selecting interesting trajectory patterns and visualizing those patterns to the user. This shapes the main focus of the visual trajectory analysis framework proposed in this paper.

Previous research of trajectory data mining and visual analytics frameworks includes the MoveMine framework [16] and the frequent trajectory pattern mining frameworks in [4, 15, 24]. Those research studies investigate to little detail the user interaction and visualizations of output trajectory patterns. As a result, the researchers developed a new framework specifically designed for interactive visual trajectory pattern mining as can be seen in figure 2.

The framework consists of 5-layers architecture. Layer 1 is the physical store of input data consisting of a Trajectory Data Warehouse (TrajDW). The TrajDW will store the details about the raw trajectories and moving-objects in a spatial database. The second layer will involve data mining patterns as an outcome of analysis techniques being executed over the TrajDW data. This is applied using data mining software tools supporting the techniques required. The techniques involve frequent sequence pattern mining (trajectory pattern mining), association rule mining and some query-based mining. Similar to research studies like [3], trajectory patterns will only focus on frequent patterns between points-of-interest (POIs) without limiting the patterns to a specific temporal timeframe. POIs consist of specific spatial locations of semantic meaning to the decision maker. This can be a shop's coordinates on a geospatial map or a fixed location point on any spatiotemporal map having relevance to the decision maker.

The third layer involves the data layer for the visualization application which stores data in a Relational Data-Base Management System (RDBMS). It stores extra thematic and demographic data about points of interest which match the different business applications' needs in the POI DB and also supports by storing visual analysis results (in data files within the Visualization Data Store). It also stores the output from the data mining of the previous phase in the Semantic Trajectory Patterns (SemTP) database. The SemTP database stores semantic trajectory patterns as defined in previous literature like [3] which represents trajectories as moves between a series of POIs.

In the fourth layer, a graphical user interface (GUI) is constructed to visualize the trajectory patterns and allow for user interaction with the Business Intelligence (BI) application (on the BI servers). Many BI tools supporting spatiotemporal visualizations are available to be used as tools for mapping the data and gathering data from the layers below. On the other hand, application logic which will handle the user-interactions and communications of the five layers is developed using BI dashboard technologies commonly used for visual analytics.

At the top-most and final fifth layer, the application user will interact with the BI application GUI and the application will process the data and return different results. The arrows between the layers means data flow and the thunderbolt from the top-most fifth layer indicates the interactive activities conducted by the decision makers.

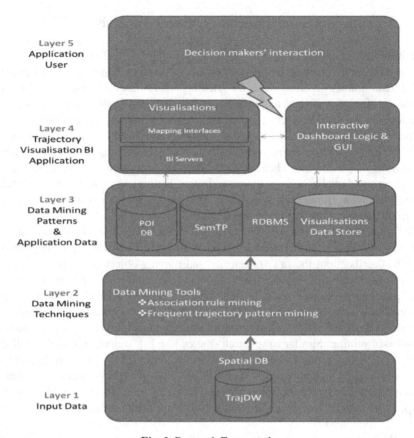

Fig. 2. Research Framework

4   Case Study: Baggage Handling Data

The framework discussed in Section 3 is applied using a prototype to a case study involving BHS baggage handling data. This involves a prototype which is tested with the main users. Research in visual analysis requires studying the perceptions and cognitions of the users of the system and the study of requirements engineering [13]. Evaluations include experimental testing with participants, panel discussions with an organisation and user testing. The tests involve the business users recording their experience and comments using real data for their own organisation. The tests will evaluate the usability and impact on decision making achieved by utilizing the prototype under research. In this case study, workshops involving panel discussions with the operational teams in the airport was executed. The operational teams include the Terminal Operations Centre (TOC) personnel responsible for managing the day-to-day operations in an airport and the BHS operations personnel.

The airport considered in this case study is a large airport which can handle more than 11,000 passengers per day in 200 scheduled international and domestic flights. This is supported by about 100 check-in counters feeding-in baggage into the BHS.

The BHS system in the airport can handle about 4800 standard pieces of baggage per hour and up-to 5 levels of baggage security screenings per bag. This kind of capacity needs to be effectively managed in order to lead to the required service-levels agreements and to maintain operational efficiency.

The framework from figure 2 was implemented in the prototype. A trajectory data warehouse based on a relational database was constructed to store details about the raw trajectories and moving-objects from the BHS departure data. This included the baggage movement data, the background semantic information (e.g. passenger information and flight information), in addition to the spatial information about the BHS (i.e. the locations and their properties).

The second layer of the framework involved data mining patterns as an outcome of analysis techniques being executed over the TrajDW data. Association rule mining and frequent trajectory pattern mining were used. Association rule mining found association rules between locations and a baggage stay duration category. Four categories based on dividing the durations into four equal quartiles were generated. An example rule would be: Screening Level-1 XRAY 2 Right → Category B (support 20%, confidence 90%). This means that the bags which stop at the level-1 screening X-RAY machine number 2 on the right section of the BHS stay for a specific duration range assigned as category B (this is explained to further details later in this section).

Frequent trajectory patterns were also found. Those are locations which are commonly visited together by a trajectory. For example a pattern like [Screening Level-1 XRAY 2 Right, Sortation 1 Right, Sortation 2 Left → Lateral 10 Left] (support 3%, confidence 20%) means that 3% of bags visit all the locations in the rule. The confidence means that 20% of the bags visiting the locations of the left hand side of the rule (Screening Level-1 XRAY 2 Right, Sortation 1 Right and Sortation 2 Left) also visit the location on the right hand side of the rule (Lateral 10 Left).

A GUI was constructed to visualize the trajectory patterns and to allow for user interaction with the BI application. The BI application logic also handles the user-interactions and communications of the five layers using data extraction and visualization techniques.

Spatial analysis output is delivered to the system user by providing map visualizations accompanied by textual results. The prototype in this research allowed for interaction with the system user. Multiple dashboards where criteria are selected and output is given on maps and charts were utilised.

The database is the core component for storing the data used in the case study consisting of BHS log files (operational data files). It stores the data physically for querying, extraction by business intelligence / reporting applications and for data mining applications. The data pre-processing includes removing non-necessary data and reformatting the XML messages from the BHS log files into a format supported by the database (spatial ANSI standards for relational databases).

The duration of the data collection was from 8th of March 2012 with the business understanding phase and until 26th of September 2012 with the completion of the final prototype evaluation with the airport personnel. The final dataset of sample data from the BHS consisted of log-file data for departure baggage. The final data sample consisted of a week's data from 3rd of July till the 9th of July 2012. The data sample was selected in July because it is one of the most active months in the year with high traffic of international travel during the summer season.

The main goals for data mining identified with the airport's BHS operations personnel and TOC personnel resulted into the main goals in table 1.

Table 1. BHS Data Mining Goals

Problem	Solution Requires
Avoiding common problems, such as congestions, within the BHS	• This can be achieved by reducing the amount of baggage going to common locations. • Better resource management during the morning / evening peak hours • Better or ideal allocation of laterals within the BHS system
Avoid uneven depreciation and unnecessary maintenance on allocated resources (e.g. conveyors) – equal distribution and loading on the left and right sides of the system	• This can be done by better or ideal allocation of laterals within the BHS system. • Better operations on the sorters and reduced bottlenecks.

A map of the BHS system modelled to-scale was also constructed to visualize the baggage trajectories spatial data. This supported the mapping visualizations of baggage trajectories in the prototype by using the map within the visualization interface.

Three iterations of modelling and visualization were executed. Each iteration involved running the data mining techniques on the data and developing the visual layers for the patterns, and then evaluations with the BHS personnel took place to assess the effectiveness of the approach and to configure the models / visualizations to meet the business needs and the requirements. This led to an effective prototype, which can help in finding BHS related problems and which can support in BHS decision-making. To handle those goals, visual results of data mining were presented to users. Interactive Human-computer interactions lead to the solution. The process is semi-automatic as the opinion of the expert guides the data mining process. Filters based on specific parameters are a common way of enhancing visualizations of large spatiotemporal datasets and were used in the visual analytics in our approach.

4.1   The Visual Analytics Techniques Used

Three visual analytics techniques were applied in this research paper and are described in table 2, including: frequent trajectory patterns within the BHS, most common BHS locations and BHS locations movement congestions (stay-durations).

Table 2 gives a title for the visualization, followed by the type of data mining (DM) applied, the description of the visualization, the description of the interaction with the user, a description of how the visualization and mining output helps the BHS decision makers and finally the evaluation of the decision makers for the visualization (result of user testing and panel discussions concerning each visualization).

Table 2. Visual Analytics Techniques in the BHS Case-Study

Visualization	Description
1. **Most common** **BHS locations** **(figure 3)**	**DM Technique:** Not applicable (direct Structured Query Language SQL querying). **Visualization Description:-** Density maps of BHS locations. - Main spatial map shows locations in the BHS as numbered boxes. - Overlay of a coloured squared box is used to indicate the congestion of the location in the BHS as a count of the number of bags. - The darker the colour of the overlaid box the more congested the location is. - The selected location types only are visualized using the shaded overlay boxes over the location. **Interaction Description:-** - Clicking a location gets an overlaid text-box giving a description of the location and the number of bags value. -Filtering by selecting the types of locations to visualize from the right side with a checkbox for each location type. - Dynamic adaption to the selection so that the colour shading considers the minimum and maximum bags count for the selected location types only. **Contribution to Decision makers:-** - Identify location movement congestions which need to be handled by BHS personnel. - Identify utilization levels of each location in the BHS by number of bags reaching it. **Evaluation of Decision Makers:-** - Easy to use interface for quickly identifying BHS locations with movement congestions, based on previous trajectory patterns left by moving baggage in the BHS.
2. **Frequent** **trajectory** **patterns from /** **to a location** **(figure 4)**	**DM Technique:** *Frequent Pattern Mining (Trajectory Pattern Mining)* **Visualization Description:-** Frequent trajectory patterns visualization consisting of map POI plots (top) and Path graphs (bottom). - On the left there are two visualizations. The top visualization shows the locations in the frequent trajectory pattern as overlaid points. The bottom shows the trajectory path with a sequential order of the locations. **Interaction Description:-** - The top two filters on the right (slider selectors) allow the user to specify the range of frequency and confidence for frequent trajectory patterns to select from using the trajectory selector towards the bottom on the right side of the visualization. - The user can select on the right side the frequent trajectory patterns to see using the start location and end location filters.

Table 2. (*continued*)

	- Clicking a BHS location visited results into an overlay of more details about the trajectory pattern. This includes the exact frequency (support), the confidence level, the trajectory (as a sequence of locations), and details describing each location. **Contribution to Decision makers:-** - Frequent trajectory patterns find common locations visited by the same bag which helps in finding the most popular trajectories between two locations / regions. **Evaluation of Decision Makers:-** - The frequent trajectory patterns can help the airport in the routing and distribution within BHS system by detecting frequent patterns of movements which need optimization within the BHS system. - The visual analytics help the BHS operations personnel look for interesting patterns with a reasonable frequency (not too low to have significance and not too high to be trivial) and for patterns between special locations of the BHS of operational significance for decision making.
3. **Location congestion category based on frequency of visits at specific timings** **(figure 5)**	**DM Technique:** *Association Rule Mining* - Finds the association of a location with a specific category of stay duration for a specific day. **Visualization Description:-** Location distribution map. - For the BHS distribution map, if we select a day (right) we will get the locations marked by points having colours indicating how long bags stayed for (i.e. the stay category associated with the location for this day). - The size of the location marks indicates the frequency of the rule. The colour indicates the category type (as seen in the legend on the right side). **Interaction Description:-** - The user can interactively select frequency (support) and confidence ranges (using the slider selectors on the right side) to see visualizations meeting those criteria only. - Clicking on the circular objects for each location show more details about the location, the stay category and the support and confidence. **Contribution to Decision makers:-** - Time Per bag at each BHS location helps identify location distribution and congestion. **Evaluation of Decision Makers:-** - Good to track average performance for complete working days. - Drawback was the unavailability of sufficient drill-down capabilities to know the reasons behind the congestion. Other visualizations showing individual trajectories had to be used to identify reasons for delays at visited locations. Better to have drill-down capabilities that show aggregate trajectory characteristics for frequent locations.

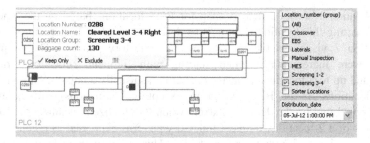

Fig. 3. Most Common BHS Locations Visualization

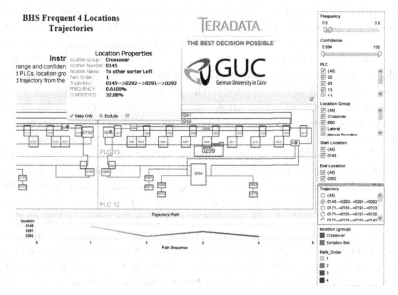

Fig. 4. Frequent Trajectory Patterns Visualization

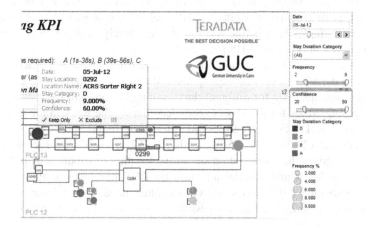

Fig. 5. Location Congestion Visualization

Visualization 1 shows a density map for BHS locations based on the amount of baggage visiting this location (figure 3) as described in table 2. On the other hand, visualization 2 visualizes the output frequent trajectory patterns. The frequent k-itemsets were incrementally found starting with frequent 2-locations trajectory patterns and up to 5-locations trajectory patterns (as usually a bag does not visit more than 5 locations in the BHS except in exceptional situations). An example of observations found by the decision makers when they utilized visualization 2 include an example of a 2-locations frequent trajectory patterns expressed as [island 4 check-in → Right side sorter] with a support of 21%. This indicates that island 4 is the most used check-in location for flights reaching the sorter on the right side of the BHS as it is the pattern with highest support for the right-hand-side of the rule. The skewness of baggage towards island 4 can mean more congestion can happen for island's 4 feedlines to the BHS, and a possible workaround this can be reallocation of check-in counters to a different check-in island. On the other hand, the confidence becomes more important for trajectory patterns on the lines for the security checks. For example, the trajectory pattern [unclear level 1-2 bags on left side → cleared level 3-4 left side] had a confidence of 23% only. This means that 77% of the bags which are unclear from levels 1-2 screening areas are most probably unclear in levels 3-4 too and require manual inspection in level 5 screening (which is manual screening and is human intensive, and having a low efficiency task that should be prevented).

Similarly, an important 3-locations frequent trajectory pattern was [sortation belt 1 right → Sortation belt 2 right →sortation belt 1 right] which had a 4% movement frequency and confidence of 24% (with the right hand side of the rule being the last location only). If you notice in the rule that the first and last location in the pattern is the same, meaning that the bag is moving in a loop on the sortation belt on the right side which has 2 belts, the ones in the rule, connected together in a circular orientation. The rule is important for BHS personnel because the bags should not go on a loop on the sortation belt. Looping movement on the sortation belts indicated lateral problems preventing the bag from being routed to its final destination of its flight's lateral. Nearly a quarter of the bags on the belt go in a loop as can be seen from the confidence measure which is not a good indicator. In addition, the frequency as a percentage of all movements in the BHS is also large enough to have operational significance for the BHS. This rule was found interactively by the BHS personnel and had severe importance for the BHS personnel.

As for visualization 3, association rule mining was used to associate between BHS locations and the amount of time bags spend at the location. The amount of time was divided into four quartiles for different duration ranges. Stay durations were divided into the equal quartiles for the stays experienced during the week in the data sample using a histogram. The Four categories (quartiles) were found as follows: A (1 second – 38 seconds), B (29 seconds – 56 seconds), C (57 seconds – 83 seconds), and D (above 83 seconds). The visualization shows the location distribution map. Locations with darker circles require special attention by BHS personnel as they are highly associated with longer duration stays. On the other hand, lighter locations indicate smooth operations which are least congested and can utilized more to balance usage.

Using visualization 3, it was found that on a daily basis there are 2 peak times (with high baggage congestion): morning peak time (biggest peak time) from 6 A.M. till 8 A.M. and another evening peak time at 8 P.M. till 11 P.M. On the other hand, some timeframes always had a low baggage flow. For example, between 12 A.M. and 3 A.M. Generally, the location congestion visualization helped find that 4 laterals (POIs) are always highly congested and need to be relaxed by assigning some of its flights to other lateral chutes which are rarely used (two such laterals were found). An important observation from the location congestion visualization also found that as time passes, along the same day from midnight till noon, the more congested the Manual Encoding Station (MES, for manual baggage barcode reading) gradually gets. This might be due to lower productivity of personnel as they move towards the end of their shift. This result can affect decision making by rearranging shifts, for example. Another example was on the 3rd of July 2012 the rule [MES right → Category D] had a confidence of 99%. This means that 99% of the bags entering the MES stay for more than 83 seconds (over a minute) which can lead to congestions and downgraded performance in the MES. This can be resolved by changing the shifts more frequently between the MES baggage handling personnel or adding more MES stations to handle the large traffic of baggage.

4.2   Discussion

Overall, the trajectory visual analytics were found to be highly effective in enhancing the level of understanding for patterns. It was also easy to use and was quoted from the decision makers to have a "low complexity".   As for decision making, the prototype helped in taking decisions about reorganization and expansions of the BHS, support in taking decisions about the lateral & counter assignments, and support in enhancing efficiency of BHS. The agility of the prototype was found to lead to faster insights using the visual analytics about the operations of the BHS. In addition, it was found that the highly customizable and flexible visualization can support exploratory investigation in the BHS operational data and baggage movement trajectory data.

Decisions supported include resource allocation. For example, lateral chutes (a location for collecting baggage per flight) in the BHS where congestions commonly take place can be allocated more baggage loading personnel to support reducing such congestions. In addition, highly congested locations (e.g. for the crossover location that links the right and left side of the BHS in the case study) can be expanded with more transportation lines (conveyors). The interactive visual analytics with the baggage movement can also help define and manage KPIs. For example, calculating the average duration of baggage in the BHS (average trajectory durations).

The most usable capabilities of the prototype were the trajectory mappings and the filtering interactions in most of the visual analytics which allow for looking for interesting patterns. The main shortcomings though that were experienced and mitigated in some of the iterations of the prototype testing was the Online Analytical Processing (OLAP) capabilities. Drill-down and roll-up OLAP capabilities to see the movements per day/week/month/year, in addition to underlying data causing some of the observations is necessary in the point of view of the panel discussion participants.

5   Conclusion and Future Work

This research paper investigated useful ways of visually and interactively analysing locations' trajectory patterns in order to help provide intelligence to the decision makers. This research also pioneered the integration of the different isolated techniques of visual analytics discussed in previous research for trajectory pattern analysis. The analytical visual approach also integrates many of the research concepts like trajectory data mining and spatial visualization which is currently still under development and need more research. The research outcomes can contribute to multiple applications and especially BHS. Bottlenecks and congestions, repetitive routes, and similar locations can be detected for supporting different decisions.

This research paper has also investigated in a case study the application of multiple visual analytics in a coherent framework. The framework had an important focus on the integration between the user, the BI layer and the data mining results. The evaluations indicate that the overall techniques used in the research framework were effective in extracting useful patterns from the trajectory data to support decision making

The evaluations also showed some visual analytics deficiencies which can be targeted in future research. Interactive drill-down capabilities within frequent trajectory patterns and locations patterns were specified to have importance for the effective visual analytics within trajectory data and can therefore be further investigated in the future.

References

1. Agrawal, R., Imieliński, T., Swami, A.: Mining Association Rules between Sets of Items in Large Databases. ACM SIGMOD Record 22(2), 207–216 (1993)
2. Al-Serafi, A., Elragal, A.: Trajectory Data Mining: a Novel Distance Measure. In: The Fifth International Conference on Advanced Geographic Information Systems, Applications, and Services GEOProcessing, pp. 125–132 (2013)
3. Alvares, L.O., Bogorny, V., de Macedo, J.A., Moelans, B., Spaccapietra, S.: Dynamic Modeling of Trajectory Patterns using Data Mining and Reverse Engineering. ER 2007 Tutorials, Posters, Panels and Industrial Contributions at the 26th International Conference on Conceptual Modeling, pp. 149–154. Australian Computer Society, Inc. (2007a)
4. Andrienko, G., Andrienko, N., Wrobel, S.: Visual analytics tools for analysis of Movement data. SIGKDD Explorations 9(2), 38–46 (2007b)
5. Andrienko, N., Andrienko, G.: Designing Visual Analytics Methods for Massive Collections of Movement Data. Cartographica: The International Journal for Geographic Information and Geovisualization 42(2), 117–138 (2007a)
6. Andrienko, N., Andrienko, G.: Visual analytics of movement: an overview of methods, tools, and procedures. In: Information Visualization, pp. 1–29 (2012)
7. Brakatsoulas, S., Pfoser, D., Tryfona, N.: Modeling, Storing and Mining Moving Object Databases. In: Proceedings of the International Database Engineering and Applications Symposium, IDEAS 2004, pp. 68–77 (2004)

8. Giannotti, F., Nanni, M., Pedreschi, D., Pinelli, F.: Trajectory Pattern Analysis for Urban Traffic. In: Proceedings of the Second International Workshop on Computational Transportation Science, IWCTS, pp. 43–47. ACM (2010)
9. Han, J., Cheng, H., Xin, D., Yan, X.: Frequent pattern mining: current status and future directions. Data Mining and Knowledge Discovery 15(1), 55–86 (2007)
10. Heinz, S.F., Pitfield, D.E.: British airways' move to Terminal 5 at London Heathrow airport: a statistical analysis of transfer baggage performance. Journal of Air Transport Management 17(2), 101–105 (2011)
11. Johnstone, M., Creighton, D., Nahavandi, S.: Status-based Routing in Baggage Handling Systems: Searching Verses Learning. IEEE Transactions on Systems, Man, and Cybernetics, Part C: Applications and Reviews 40(2), 189–200 (2010)
12. Kapler, T., Wright, W.: GeoTime Information Visualization. In: Proceedings of the IEEE Symposium on Information Visualization, pp. 25–32 (2004)
13. Keim, D., Andrienko, G., Fekete, J.-D., Görg, C., Kohlhammer, J., Melançon, G.: Visual Analytics: Definition, Process, and Challenges. In: Information Visualization, pp. 154–175 (2008a)
14. Lee, S.J., Siau, K.: A review of data mining techniques. Industrial Management & Data Systems 101(1), 41–46 (2001)
15. Leonardi, L., Marketos, G., Frentzos, E., Giatrakos, N., Orlando, S., Pelekis, N., et al.: T-Warehouse: Visual OLAP Analysis on Trajectory Data. In: Proceedings of the IEEE 26th International Conference on Data Engineering, ICDE, pp. 1141–1144 (2010)
16. Li, Z., Ji, M., Lee, J.-G., Tang, L.-A., Yu, Y., Han, J., et al.: MoveMine: Mining Moving Object Databases. In: Proceedings of the 2010 ACM SIGMOD International Conference on Management of Data, pp. 1203–1206 (2010)
17. Pelekis, N., Frentzos, E., Giatrakos, N., Theodoridis, Y.: HERMES: Aggregative LBS via a Trajectory DB Engine. In: Proceedings of the 2008 ACM SIGMOD International Conference on Management of Data, pp. 1255–1258 (2008)
18. Rizk, A., Elragal, A.: Trajectory Data Analysis in Support of Understanding Movement Patterns: A Data Mining Approach. In: Proceedings of the Eighteenth Americas Conference on Information Systems, AMCIS, pp. 1–8 (2012)
19. Samola, B.: Innovations in passenger and baggage processing at Schiphol Airport. Journal of Airport Management 2(3), 227–234 (2008)
20. Schreck, T., Bernard, J., Tekušová, T., Kohlhammer, J.: Visual Cluster Analysis of Trajectory Data With Interactive Kohonen Maps. In: Proceedings of the IEEE Symposium on Visual Analytics Science and Technology 2008, VAST, pp. 3–10 (2008)
21. SITA, Baggage Report 2012 (2012), SITA website: http://www.sita.aero/content/baggage-report-2012 (retrieved May 1, 2012)
22. Spaccapietra, S., Parent, C., Damiani, M.L., de Macedo, J.A., Porto, F., Vangenot, C.: A conceptual view on trajectories. Data & Knowledge Engineering 65(1), 126–146 (2008)
23. Wang, J.-B., Fan, C.-J., Fu, H.-G.: Discussion on Airport Business Intelligence System Architecture. International Journal of Business and Social Science 3(13), 134–138 (2012)
24. Yan, Z., Chakraborty, D., Parent, C., Spaccapietra, S., Aberer, K.: SeMiTri: A Framework for Semantic Annotation of Heterogeneous Trajectories. In: Proceedings of the 14th International Conference on Extending Database Technology, EDBT/ICDT, pp. 259–270. ACM (2011)
25. Yuan, J., Zheng, Y., Sun, G.: T-Drive: Driving Directions Based on Taxi Trajectories. In: Proceedings of the 18th SIGSPATIAL International Conference on Advances in Geographic Information Systems, ACM GIS, pp. 99–108 (2010)
26. Zhao, Q., Bhowmick, S.S.: Sequential Pattern Mining: A Survey. Technical Report, Nanyang Technological University, CAIS, Singapore (2003)

Analysis of Trajectory Data in Support of Traffic Management: A Data Mining Approach

Ahmed Elragal[1] and Hisham Raslan[2]

[1] Department of Business Informatics & Operations, German University in Cairo (GUC),
Cairo, Egypt
ahmed.elragal@guc.edu.eg
[2] Teradata Egypt,21 Giza st, Giza, Egypt
hisham.raslan@teradata.com

Abstract. Huge amount of location and tracking data is gathered by location and tracking technologies, such as global positioning system (GPS) and global system for mobile communication (GSM) devices; leading to the collection of large spatiotemporal datasets and to the opportunity of discovering usable knowledge about movement behavior. Movement behavior can be extremely useful in many ways when applied, for example, in the domain of traffic management, planning metropolitan areas, mobile marketing, tourism, etc. In this research, we move towards this direction and propose a framework for finding trajectory patterns of frequent behaviors using GSM data. The research question is "how to use trajectory data analysis in support of solving traffic management problems utilizing data mining techniques?" Our framework is illustrated to explain how GSM data can provide accurate information about population movement behavior, and hence support traffic decisions.

Keywords: trajectory, data mining, traffic management.

1 Introduction

The movement of people or vehicles within a given area can be observed from the digital traces left behind by the personal or vehicular mobile devices, and collected by the wireless network infrastructures. For instance, mobile phones leave positioning logs, which specify their localization, or cell, at each moment they are connected to the GSM network. The increasing use of these technologies will make available large amounts of data pertaining to individual trajectories; therefore, there is an opportunity to discover, from these trajectories, spatiotemporal patterns that convey useful knowledge. Spatiotemporal patterns that show the cumulative behavior of a population of moving objects are useful abstractions to understand population movement behavior. In particular, a form of pattern, which represents an aggregated abstraction of many individual trajectories of moving objects within an observed population, would be extremely useful in the domain of sustainable mobility and traffic management in metropolitan areas, where the discovery of traffic flows among sequences of different places in a town can help decision makers take well informed decisions in different areas such as traffic management and urban planning.

P. Perner (Ed.): ICDM 2014, LNAI 8557, pp. 174–188, 2014.

In many application domains, useful information can be extracted from moving object data if the meaning as well as the background information are considered. The knowledge of moving patterns between different places in the geographic space may help the user to answer queries about moving objects or movement behavior. In order to capture and model such pattern relationships, data mining techniques play an essential role.

For the purpose of this research, traffic management is the direction, control, and supervision of all functions related to road-related passenger transportation services. Traffic management is a key towards the creation of a modern and sustainable city. Traffic management becomes a rather critical job in an over-populous city with limited resources. In a city over populated like Cairo with millions of vehicles and very limited roadways, traffic management is impossible without the aid of technology.

The goal of this research is to develop framework for finding patterns in trajectories utilizing data mining techniques. The framework can be used to find hidden patterns in trajectory data to support spatiotemporal semantic decisions in the area of traffic management and analysis. The research relies on a case study based on GSM data from Cairo to support answering the research question "how to use trajectory data analysis in support of solving traffic management problems utilizing data mining techniques?"

Our approach is directed towards building a framework to support the management and analysis of traffic data. Fig. 1. Initial framework explains the suggested framework which we are going to build and test.

Fig. 1. Initial framework

Sample data was used in the experimental case study. It's a real data taken from a GSM operator in Egypt. The data is mapped into known regions and stored in a central data warehouse where knowledge discovery tasks take place to extract useful knowledge in support of traffic management and analysis.

The remaining of this paper organized as follows: section 2, related work; section 3, traffic management; section 4 GSM data in support of traffic management; section 5, proposed framework; section 6, conclusion; and references at the end.

2   Related Work

The literature of trajectory data mining and mining moving objects has witnessed different waves focusing on various themes ranging from those focused on frameworks to those focusing on visual interaction between solution and user.

Many research focused on location information, sequence, and regions of interests e.g., [1], [2], [3], [4], [5], and [6]. [1] propose a spatial context model, which deals with the location prediction of mobile users. The model is used for the classification of the users' trajectories through Machine Learning (ML) algorithms. Predicting spatial context is treated through supervised learning. [2] Developed an extension of the sequential pattern mining paradigm that analyzes the trajectories of moving objects. They introduced trajectory patterns as a descriptions of frequent behaviors, in terms of both space (i.e., the regions of space visited during movements) and time (i.e., the duration of movements). [3] Provided a case study of movement data analysis where statistical means and pattern mining were merged together with visualization techniques to improve understanding of data. [4] Tried to cope with the complexity of trajectory semantics in terms of three evolving steps: Trajectory Modeling which considers spatiotemporal features (like trajectory modeling data type moving point) and semantic trajectory units (like stops and moves); Trajectory Computing Propose corresponding bottom up computational solutions for targeting semantic trajectories; Trajectory Pattern Discovery Investigate the mining and learning algorithms for the computed semantic trajectories. [5] Proposed a model for trajectory patterns and a measure to represent the expected occurrences of a pattern in a set of imprecise trajectories. The concept of pattern groups is introduced to present the trajectory patterns and a new min-max property was identified. A TrajPattern algorithm was devised based on the newly discovered property and the algorithm was applied on a wide range of real and synthetic data sets to demonstrate the usefulness, efficiency, and scalability of this approach. [6] Mined interesting locations and travel sequence in a given Geo-spatial region using GPS trajectories. In their paper they improved location based services by integrating Social Networking into Mobile Web.

Another wave focused on developing framework on trajectories analysis and pattern detection e.g., [7], and [8]. [7] Proposed a reverse engineering framework for mining and modeling semantic trajectory patterns. They applied data mining to extract general trajectory patterns, and through a new kind of relationships, they model these patterns in the geographic database schema. They used a case study able to shows the power of the framework for modeling semantic trajectory patterns in the geographic space. [8] developed MoveMine which is able to integrate data mining functions including moving object pattern mining and trajectory mining based on novel methods. MoveMine is able to perform trajectory clustering, classification and outlier detection. The output of the system could be written in Google maps and Google earth. Their system consists of three layers: 1- data collection and cleaning, 2- Mining, 3- Visualization interface.

Another research defined trajectory data warehouse (TDW) that is loaded by spatiotemporal observations and studied how standard data warehousing tools can be used to store trajectories and to compute OLAP operations over them e.g., [9], [10], [11]. [9] Presented an approach for storing and aggregating spatio-temporal patterns by using a Trajectory Data Warehouse (TDW). [10] Investigated the extension of Data Warehousing and data mining technology so as to be applicable on mobility data. In his work, he presented the developed framework for analyzing mobility data and some preliminary results. [11] Aims was to make trajectories as a first class concept in the trajectory data conceptual model and to design a TDW, in which data resulting from mobile information collectors' trajectory are gathered. These data will be analyzed, according to trajectory characteristics, for decision making purposes, such as new products commercialization, new commerce implementation, etc.

Further research focused on the techniques and algorithms needed to find patterns and develop solutions e.g., [12], [13], and [2]. [12] Proposed a data preprocessing model to add semantic information to trajectories in order to facilitate trajectory data analysis in different application domains. [13] Described the analysis, pre-processing, modeling, and storage techniques for trajectory data that constitute a Moving Object Database (MOD). MOD is the backbone of the 'PATH-FINDER' system, which specifically focuses on extracting further information about the movement of vehicles in the Athens municipal area.

Few research studies were directed towards visual interaction and security e.g., [14], [15], and [16]. [16] Proposed a visual-interactive monitoring and control framework extending the basic Self-Organizing Map (SOM) algorithm. The framework implements the general Visual Analytics idea to combine automatic data analysis with human expert supervision. It provides facilities for visually monitoring and interactively controlling the trajectory clustering process. They Applied the framework on a trajectory clustering problem, to demonstrate its potential in combining both unsupervised (machine) and supervised (human expert) processing, to produce appropriate cluster results. [14] Proposed a general framework to solve the conflict between the data mining methods, which want as precise data as possible, and the users who want to protect their privacy by not disclosing their exact movements. The framework allows user location data to be anonymized, thus preserving privacy, while still allowing interesting patterns to be discovered. The framework allows users to specify individual desired levels of privacy that the data collection and mining system will then meet. [15] Studied the privacy threats in trajectory data publishing and show that traditional anonymization methods are not applicable for trajectory data due to its challenging properties: high-dimensional, sparse, and sequential. Their primary contributions are (1) to propose a new privacy model called LKC-privacy that overcomes these challenges, and (2) to develop an efficient anonymization algorithm to achieve LKC-privacy while preserving the information utility for trajectory pattern mining.

3   Traffic Management

In recent years, urban traffic congestion has become a huge problem in many cities across many countries. In order to reduce congestion, governments usually invest in improving city infrastructures. However, infrastructure improvements are very costly to undertake; hence, existing infrastructure and vehicles have to be used more efficiently. To reduce traffic congestion, it is necessary to conduct further research on the various characteristics of traffic flow patterns. In general, road traffic system consists of many autonomous, such as vehicle users, public transportation systems, traffic lights and traffic management center, which distribute over a large area and interact with one another to achieve an individual goal. Traffic management objective is to increase the efficient passages of every vehicle, while at the same time reduce the number of vehicles on the street. Therefore, research on traffic information control and traffic guidance strategies are particularly necessary and important [17].

Urban traffic analysis and control is a complex problem that is difficult to analyze with traditional analytical methods. The degree of complexity of vehicle movement in urban centers is such that modeling and simulation techniques have been gaining popularity as analysis tool. Simulation entitles the study of particular problems, allowing providing solutions based on experimentation. [18] Presented the results of a project to build modeling and simulation tools with this purpose. The first stage of this project was devoted to define and validate a high level specification language representing city sections. This language, called ATLAS (Advanced Traffic LAnguage Specifications) focuses on the detailed specification of traffic behavior. ATLAS is a specification language defined to outline city sections as cell spaces. A static view of the city section to be analyzed can be defined and a modeler is able to define complex traffic models in a simple fashion. [19] Used the term 'urban data-mining' which they described as a methodological approach that discovers logical or mathematical and partly complex descriptions of urban patterns and regularities inside the data.

Trying to understand, manage and predict the traffic phenomenon in a city is both interesting and useful. For instance, city authorities, by studying the traffic flow, would be able to improve traffic conditions, arrange the construction of new roads, the extension of existing ones, and the placement of traffic lights. This target can be served by analyzing traffic data to monitor the traffic flow and thus to discover traffic related patterns. These patterns can be expressed through relationships among the road segments of the city network. We aim to discover, by using aggregated mobility data, how the traffic flows in this network, the road segments that contribute to the flow and how this happens. We believe that, the application of new information technologies such as trajectory mining technology to urban traffic information control can make it possible to create and deploy more intelligent systems for traffic control and management to support road managers in traffic management tasks.

4   GSM Data in Support of Traffic Management

Traffic management is analyzed and studied either based on original data collected by means of cameras, GPS, or other tracking systems. In the absence of this, traffic management in studied using synthetic data. However, synthetic data does not really represent the real problem and is hard to generalize. Meanwhile, collecting traffic data needs certain setup, which is not always available. Therefore, in our framework we used GSM data to understand the traffic patterns and behavior. In the following we will describe how GSM data can help for that purpose.

GSM mobile network is a radio network distributed over land areas called cells, each served by at least one fixed-location transceiver, known as a cell site or base station. When joined together these cells provide radio coverage over a wide geographic area. This enables a large number of portable transceivers (e.g., mobile phones, pagers, etc.) to communicate with each other and with fixed transceivers and telephones anywhere in the network, via base stations. The most common example of a cellular network is a mobile phone (cell phone) network. A mobile phone is a portable telephone which receives or makes calls through a cell site (base station), or transmitting tower. As the phone user moves from one cell area to another cell whilst a call is in progress, the mobile station will search for a new channel to attach to in order not to drop the call. Once a new channel is found, the network will command the mobile unit to switch to the new channel and at the same time switch the call onto the new channel.

Every time a subscriber makes or receives a call the network generates a Call Details Record CDR to store the details of the call (caller ID, called ID, time of the call, call duration, etc.) one of the CDR parameters is the cell id. It is possible to follow the path (trajectory) of a subscriber by linking the CDRs generated by the subscriber in a period of time. To study the traffic in certain areas over specific roads, we need to identify the cell ID's used by the mobile operator to cover those areas then map their locations to geographical areas (semantics) as shown in Fig. 2. Cells mapped on traffic roads. Using the CDRs generated with the identified cells over a period of time we can identify moving pattern of the operator subscribers in these areas.

Fig. 2. Cells mapped on traffic roads

5   The Proposed Framework

The proposed framework aim is to establish an enhanced approach towards building analysis engine that can detect movement pattern that if known to the traffic department, is expected to influence their decisions and hence enhance traffic quality and save time, fuel consumption, and thereafter help Cairo becomes an environment friendly city. Based on the conducted business analysis, the traffic department is interested in analyzing data in specific areas such as traffic volume and Traffic pattern on specific roads.

The framework includes process, technologies, data, and decisions as the main aspects towards knowledge extraction. The process outlines the detailed tasks to take place at each phase and highlights the main layers which data passes through to be transformed into knowledge. The technologies used in each layer or by the tasks are also presented. The phases of the proposed framework are presented in Fig. 3. The proposed framework. The following are the tasks performed to reach the results:

1- Build Logical Data Model
2- Data loading
   • Load CDRs
   • Load Cell information
   • Create and load lookups
3- Data preparation & Semantic annotation
   • Data Quality assessment
   • Define POI to add semantics to the data
   • Identify Commuters (users)
4- Data Preprocessing
   • Trajectory Extraction
   • Build ADS
5- Perform Analytics.

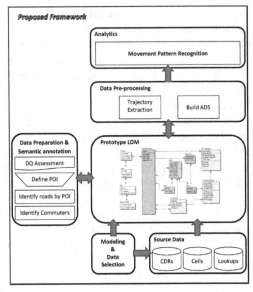

Fig. 3. The proposed framework

While using GSM data to study traffic pattern, we will not be violating people's privacy i.e., we will be using disguised data, which does not reveal real identify of people.

In the following we will provide description of each task.

5.1   Build the Logical Data Model

The foundation for the analysis engine is a well-designed Logical Data Model. The model will be used to physically realize the engine and will help the analytical users plan and develop queries and analytics. Fig. 4. LDM shows the proposed LDM that will be used to build analysis engine.

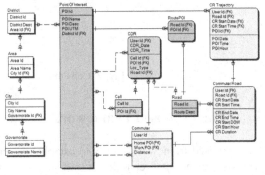

Fig. 4. LDM

The model is comprised of entities, attributes and relations; the model entities are colored to differentiate the type of information stored in each entity; blue entities are data loaded from the source (CDRs and Cells), white (uncolored) entities have semantics information (lookups: governorate, cities, areas, districts), green represents entities identified for analysis (POIs and roads), and yellow entities contain information extracted using analysis algorithms (trajectories).

5.2   Data Loading

Data loading includes the following tasks:
1. Load CDRs and Cell information
2. Create and load lookups

Load CDRs and Cell Information

Call Detail Record tables for voice and GPRS are merged into the table "CDR". The CDR table has the following structure: (User_Id, CDR_Date, CDR_Time, Cell_Id)

- Four Months of voice and data CDR's are merged into the table "CDR". Total number of CDRs is 10,314,009,634 ≈ 1.5B
- Greater Cairo cells loaded to "Cell" table in the cell id column; the POI column will be mapped later. Total number of cells is 14,175

After loading the CDR table, cell ids are loaded to the "Cell" table in the cell id column; the POI column will be mapped later. POI's were manually identified using Google earth by locating all cells on the map then grouping the cells in a specific area to a POI as we will explain later. The cell table has the following structure: (Cell_Id, POI_Id)

Create and Load Lookups

Lookups tables Governorate, City, Area, and District are created and loaded with the information required to add semantics to the analysis results.

5.3   Data Preparation and Semantic Annotation

Data Quality Profiling

Data quality is the suitability of data to meet business requirements. Because different organizations and applications have different uses and requirements for the data, data quality requirements will also differ. So data doesn't have to be perfect, but it needs to meet business requirements. We will be assessing the data for Consistency (Format and Content), Completeness, Uniqueness, and Integrity

We use data profiling tool to perform the required data quality assessments. The selected tool is part of Teradata Warehouse Miner TWM. The following tables, Table 1. Value analysis for the CDRs table, and Table 2. Value analysis for the cells table, show value analysis for the CDR and Cell table respectively.

Table 1. Value analysis for the CDRs table

Column Name	Count	null	Unique	Zero	Positive	Negative
User_ID	10314009634	0	15424023	0	10314009634	0
CDR_Date	10314009634	0	122			
CDR_Time	10314009634	0	86400	138790	10313870844	0
Cell_ID	10314009634	0	15077	0	10314009634	0

Table 2. Value analysis for the cells table

Column Name	Count	null	Unique	Zero	Positive	Negative
Cell_ID	14175	0	14175	0	14175	0
Cell_Status	14175	151	4			
POI_Id	14175	0	522	0	14175	0

Data Issues
1. Number of CDRs with missing cells: 214026321 (2% of total CDRs)
2. Number of missing cells: 1127 (8% of total Cells)

The ratios of discovered data issues (2%) will not affect the accuracy of the analysis results and can be ignored

Define Points of Interest POI
This step is comprised of the following:
1. Group cell sites to define Points of Interest POI
2. Update POI column in the cell and CDR tables

Group Cell Sites to Define Points of Interest POI
Points of Interest POI under investigation are defined by nearest cell sites. We also use the cell sites on the roads to monitor traffic events; such points are also considered POI. To locate cell sites, all cells were added to Google earth map and the cell sites near by the roads were selected (manually) and defined as POI. To add the cells to the map we use the site coordinates. The coordinates are in Universal Transverse Mercator notation (UTM).

Cell sites usually have more than one cell; also the number of cells is very large to handle! (See Fig. 5. Cell sites located on greater Cairo map).

Fig. 5. Cell sites located on greater Cairo map

To reduce the number of cells we grouped the cells in one district to one or more points on the map in the center of the grouped cells. These points are used in our research as Points of Interest POI (see Fig. 6. POIs located on greater Cairo map).

Fig. 6. POIs located on greater Cairo map

The process of grouping cells to a centralized POI is a manual task, and also judgmental. We tried to choose the POIs to be close to a real POI and of course identified all possible POIs close to the main roads to be able to track movements on the roads. Using this process we reduced the number of points on the map from 14K to about 500 point which is more manageable and easier to locate on the map. The identified POIs and the POI demographics are then loaded in the POI table.

Update POI Column in the Cell and CDR Tables
As a result of grouping cells to POI a mapping table is created. The mapping table is used to populate the POI column in the cell table and by joining the cells table and the CDR's table on the cell id column we can update POI column in the CDR table.

Identify Roads Using POI
Roads are part of Regions of Interest that are defined by the business users. For the purpose of the study some roads are selected to capture traffic volume and study movement patterns. The selected roads are stored in the road table. Each road is identified by POIs close to the road and a mapping table that relates a road to a group of POIs is used to populate the RoutePOI table.

Identify Commuters
We mean by commuters, the users of the mobile devices who generate voice and data CDRs. To maintain the privacy of the subscribers we only have anonymous identification number (user id) for the subscribers in the CDR table. We can use user id to select the CDRs made by each user (subscriber) and use it for the purpose of this research. To populate the commuter table we select the distinct users from the CDR table and load them in the commuter table. The number of commuters (users) identified is 13,887,256 which is a considerable number of commuters, relative to greater Cairo population, that can fairly represent greater Cairo movement behavior.

5.4   Data Preprocessing

The following are the algorithms used to extract user trajectories and the Analytical Data Set ADS that will be used for the analytics part.

Trajectory Extraction
In this task we will extract the movement trajectories on specific roads from the CDRs. In this task the movement of the commuters (the required trajectories) are inferred from the CDR's generated on the specified roads. The results are stored, as shown in the logical model, in two tables, "Commuter Road" and "CR Trajectory". The "Commuter Road" stores all the trajectory a specific user has done on any road. While, the details or the stops on the road are stored in the "CR Trajectory" table.

A state diagram that explains the methodology used to detect trajectories by following the usage of a specific user in a specific period is shown in Fig. 7. State diagram for detecting movements from CDRs, followed by the algorithm used to populate the two tables. The CDR table is sorted by user_id and time stamp as the algorithm is built to detect the trajectories user by user.

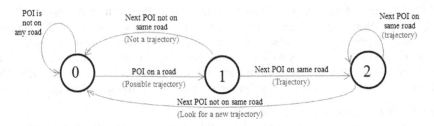

Fig. 7. State diagram for detecting movements from CDRs

```
Start
Position pointer to first row in CDR table
Loop1: Do while not EOF
  SET Luser_id = user_id;
  SET first_row_of_the_user = T;
  Loop2: Do while user_id = Luser_id
    IF First_row_of_the_user = T THEN
      SET LRoute_Id = Route_Id, store 1st row values;
      SET First_row_of_the_user = F;
      SET State = 1;
    Else
      IF LRoute_Id = Route_Id THEN
        IF state = 1
          INSERT new row in CR_Traj using 1st row values
          SET State = 2;
        End IF
        INSERT new row in CR_Traj using this row values
        Store this row attribute values as last row
      Else
        IF State = 2 THEN
          INSERT a new row in Com_Rd using first and last
        END IF;
        SET LRoute_Id = Route_Id;
        SET State = 1;
      END IF;
    END IF;
    Skip to next row
  END Loop2;
END Loop1;
```

Build ADS

To study the pattern of movement, we aggregate the trajectories per route per day of week per hour of the day. The result is the Analytical Data Set ADS used for performing the analytics in this area. {Route_id, CR_Start_DOW, CR_Start_Hour, CR_Rt_DOW_Hr_Cnt}.

5.5   Performing Analytics and Sample of the Case Study Results

Traffic Pattern

Identifying traffic patterns could be the most valuable information for traffic management; we understand that traffic pattern varies (day and night, hour of day, day of week, summer and winter, vacation times, etc.) The analysis engine should be able to support analysis over different time dimensions. For example, the day pattern is of importance to be known as in the following business questions:

BQ#1 - The night pattern (after 8PM, to 6AM); per road, per day
BQ#2 - The daytime pattern, per road, per day

The ADS was used to explore the different movement behaviors on a specific road and on all roads. Microstrategy was used to analyze the data and tables and the analysis results and graphs shown in shown in Fig. 8. Traffic pattern - all selected roads, Table 3. Daytime movement (6am – 8pm), and Table 4. Night movement (8pm – 6am). Again, busiest road is highlighted in yellow in the road row; while busiest day is highlighted in yellow in the totals row.

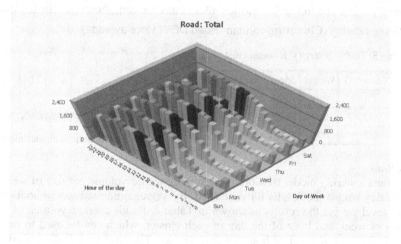

Fig. 8. Traffic pattern - all selected roads

Table 3. Daytime movement (6am – 8pm)

Road Name	Sun	Mon	Tue	Wed	Thu	Fri	Sat
26th of July axis	2,029	2,053	2,052	2,059	2,034	1,310	1,689
6th of October bridge	2,241	2,386	2,368	2,439	2,293	1,133	1,740
Salah Salem	2,199	2,237	2,225	2,241	2,161	1,168	1,656
Nasr road	3,525	3,358	3,396	3,374	3,349	1,931	2,868
Ring road – north	3,609	3,523	3,493	3,604	3,695	2,599	3,099
Ring road – south	5,170	5,115	5,128	5,114	5,125	3,052	4,197
Suez road	818	823	831	837	818	474	647
Ismailia road	1,770	1,880	1,822	1,855	1,776	1,140	1,514
Cairo-Alex Agg rd	2,506	2,556	2,522	2,556	2,640	1,558	1,999
Cairo-Alex Desert rd	1,151	1,169	1,178	1,211	1,265	657	830
	25,018	25,100	25,015	25,290	25,156	15,022	20,239

Table 4. Night movement (8pm – 6am)

Road Name	Sun	Mon	Tue	Wed	Thu	Fri	Sat
26th of July axis	614	614	660	630	617	599	619
6th of October bridge	616	657	675	674	635	566	601
Salah Salem	760	764	778	773	754	694	703
Nasr road	1,305	1,303	1,353	1,296	1,299	1,158	1,184
Ring road – north	1,476	1,483	1,522	1,518	1,458	1,279	1,333
Ring road – south	1,850	1,852	1,908	1,860	1,805	1,645	1,683
Suez road	239	244	240	232	219	233	217
Ismailia road	538	538	540	527	509	455	462
Cairo-Alex Agg rd	888	872	907	912	905	747	799
Cairo-Alex Desert rd	288	281	300	300	316	265	264
	8,574	8,608	8,883	8,722	8,517	7,641	7,865

Ranking Roads Using Clustering

Further analysis was done to cluster the movement pattern on the road using TWM data mining tool, the result of clustering the ADS to 3 clusters is listed in Table 5. Traffic pattern – K-mean clustering. The number of clusters was chosen heuristically as per common perception to traffic it is low, medium, or high. Indeed different other techniques could also be used. Additionally, different interpretations could be generated. The graph in Fig. 9. Traffic pattern cluster mean, shows the cluster means.

Clustering results: (Clustering column: Road DOW Hour average)

Table 5. Traffic pattern – K-mean clustering

Cluster ID	Weight	Mean	Variance
1	0.49	45.56	793.79
2	0.37	164.32	1299.34
3	0.15	326.72	3895.40

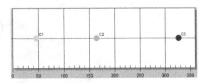

Fig. 9. Traffic pattern cluster mean

Road Rank

Using the clustering model to score road average traffic volume per day of week and hour of day we get the results for each road. By aggregating road scored hours to the cluster level we get the results as shown in Table 6. Roads average volume of traffic per day of week and hour of the day in each cluster, which can be used to rank the road based on the number of hours in each cluster. The road rank is illustrated using the colors green, yellow, and red.

Table 6. Roads average volume of traffic per day of week and hour of the day in each cluster

Road Name	Number of hours per cluster		
	1	2	3
26th of July axis	71	97	
6th of October bridge	71	97	
Salah Salem	68	100	
Nasr road	49	50	69
Ring road – north	43	57	68
Ring road – south	43	19	106
Suez road	168		
Ismailia road	93	75	
Cairo-Alex Agg rd	61	104	3
Cairo-Alex Desert rd	151	17	

6   Conclusion

In this research, we have proposed a framework, which would enable the analysis of trajectory data using data mining techniques. While mainstream literature focus on finding this knowledge based on GPS data, we have been able to show how this could be achieved based on GSM data. Traffic management decision makers could use our framework to make related decision e.g., traffic volume on specific roads, and traffic pattern. Our analyses confirm that long-term GSM activity data is well suited to identify typical movement patterns done by communities especially when GPS data is not available. In addition, we believe our methods explained how that estimation of movement quantities from GSM activity data is possible.

Using real data in the case study was definitely for the benefit of the research; however, the data size was a major obstacle that caused the research to halt several times before Teradata granted the use of one of its servers to the research. This is very important to mention as GSM CDRs are always huge in volume and for deployment the required volume of data will be much more as data from all operators should be integrated to provide complete view for whole population.

CDRs were used to extract commuter's trajectories. However, CDRs are not generated unless the commuters make network activity (call, SMS, data, etc.). In other words, not all trajectories of the commuters traveling on the roads are extracted, only trajectories for commuters that used their devices while traveling. Moreover, the commuter should make more than one network activity on the road with different cell id to be considered as a traveler in order to distinguish the moving commuter from a resident who lives close to the road. To overcome this problem, other GSM generated records can be used such as the cell registration record that is generated automatically every time a user moves between cells; however, the data size is much more than the usage CDRs. We need to take into consideration that the use of data services is increasing as a trend while voice calls are decreasing and data services generate a lot of CDRs without the user intervention (e.g. check mails and messages, perform software updates, etc.) which makes data CDRs a viable alternative in the near future. In all cases, we believe this area still needs more investigation to enhance the intelligence of the trajectory extraction algorithm.

The complete deployment of the framework by traffic department or concerned government agencies, data from the three mobile networks operating in Egypt should be integrated to get movement pattern of the whole population.

Future work includes the full deployment of the framework and its application in different business domains. Also, comparative study between GSM-based analysis versus GPS-based.

References

[1] Anagnostopoulos, T., Anagnostopoulos, C., Hadjiefthymiades, S., Kyriakakos, M., Kalousis, A.: Predicting the Location of Mobile Users: A Machine Learning Approach. In: ICPS 2009, London (2009)

[2] Giannotti, F., Nanni, M., Pedreschi, D., Pinelli, F.: Trajectory Pattern Mining. In: KDD 2007, California (2007)

[3] Giannotti, F., Nanni, M., Pedreschi, D., Pinelli, F.: Trajectory Pattern Analysis for Urban Traffic. In: IWCTS 2009, Seattle (2009)

[4] Yan, Z., Parent, C., Spaccapietra, S., Chakraborty, D.: A Hybrid Model and Computing Platform for Spatio-semantic Trajectories. In: Aroyo, L., Antoniou, G., Hyvönen, E., ten Teije, A., Stuckenschmidt, H., Cabral, L., Tudorache, T. (eds.) ESWC 2010, Part I. LNCS, vol. 6088, pp. 60–75. Springer, Heidelberg (2010)

[5] Yang, J., Hu, M.: TrajPattern: Mining Sequential Patterns from Imprecise Trajectories of Mobile Objects. In: Ioannidis, Y., et al. (eds.) EDBT 2006. LNCS, vol. 3896, pp. 664–681. Springer, Heidelberg (2006)

[6] Zheng, Y., Zhang, L., Xie, X., Ma, W.-Y.: Mining Interesting Locations and Travel Sequences from GPS Trajectories. In: WWW 2009, Madrid (2009)

[7] Alvares, L.O., Bogorny, V., de Macedo, J.A.F., Moelans, B., Spaccapietra, S.: Dynamic Modeling of Trajectory Patterns using Data Mining and Reverse Engineering. In: ER 2007, Aukland (2007)

[8] Li, Z., Ji, M., Lee, J.-G., Tang, L.-A., Yu, Y., Han, J., Kays, R.: MoveMine: Mining Moving Object Databases. In: Proceedings of the ACM SIGMOD Conference, Indianapolis (2010)

[9] Leonardi, L., Orlando, S., Raffaetà, A., Roncato, A., Silvestri, C.: Frequent Spatio-Temporal Patterns in Trajectory Data Warehouses. In: SAC 2009, Honolulu (2009)

[10] Marketos, G.: Mobility Data Warehousing and Mining. In: VLDB 2009, Lyon (2009)

[11] Oueslati, W., Akaichi, J.: Mobile Information Collectors Trajectory Data Warehouse Design. International Journal of Managing Information Technology (IJMIT) 2010 (2010)

[12] Alvares, L.O., Bogorny, V., Kuijpers, B., Fernandes, J.A., Moelans, B., Vaisman, A.: A Model for Enriching Trajectories with Semantic Geographical Information. In: GIS 2007, Seattle (2007)

[13] Brakatsoulas, S., Pfoser, D., Tryfona, N.: Modeling, Storing and Mining Moving Object Databases. In: International Database Engineering and Applications Symposium (IDEAS 2004), Coimbra (2004)

[14] Gidófalvi, G., Huang, X., Pedersen, T.B.: Privacy–Preserving Trajectory Collection. In: ACM GIS 2008, Irvine (2008)

[15] Mohammed, N., Fung, B.C.M., Debbabi, M.: Walking in the Crowd: Anonymizing Trajectory Data for Pattern Analysis. In: CIKM 2009, Hong Kong (2009)

[16] Schreck, T., Bernard, J., Tekusova, T., Kohlhammer, J.: Visual Cluster Analysis of Trajectory Data With Interactive Kohonen Maps. In: VAST 2008, Columbus (2008)

[17] Jin, X., Itmi, M., Abdulrab, H.: A Cooperative Multi-agent System Simulation Model for Urban Traffic Intelligent Control. In: SCSC 2007 (2007)

[18] Tártaro, M.L., Wainer, G.: Defining Models of Urban Traffic Using the TSC Tool. In: Proceedings of the 2001 Winter Simulation Conference (2001)

[19] Behnisch, M., Ultsch, A.: Urban data-mining: spatiotemporal exploration of multidimensional data. Building Research & Information, 520–532 (2009)

Detecting the Transition Stage of Cells and Cell Parts by Prototype-Based Classification

Petra Perner

Institute of Computer Vision and Applied Computer Sciences, IBaI
PSF 301114, 04251 Leipzig, Germany

Abstract. Unsupervised classification is the choice when knowledge about the class numbers and the class properties is missing. However, using clustering might not lead to the correct class and needs interacting with the domain experts to figure out the classes that make sense for the respective domain. We propose to use a prototype-based learning and classification method in order to figure out the right number of classes and the class description. An expert might start with picking out a prototypical image or object for the class he is expecting. Later on, he might pick out some more prototypes that might represent the variance of the class. By doing so might be incrementally learnt the class border and the knowledge about the class. It does not need the expert so heavy interaction with the system. Such a method is especially useful when the domain has very noisy objects and images. We present in the paper the method for prototype-based classification, the methodology, and describe the success of the method on a biological application - the detection of different dynamic signatures of mitochondrial movement.

Keywords: Index Terms— Mitochondrial Movement, Cell Biology, Prototype-Based Classification, Knowledge Acquisition, Class Discovery, Discovery of Class Description, Feature Selection, Prototype Selection.

1 Introduction

Prototypical classifiers have been successfully studied for medical applications by Schmidt and Gierl [1], Perner [2] for image interpretation and by Nilsson and Funk [3] on time-series data. The simple nearest-neighbor-approach [4] as well as hierarchical indexing and retrieval methods [5] have been applied to the problem. It has been shown that an initial reasoning system could be built up based on prototypical cases. The systems are useful in practice and can acquire new cases for further reasoning [5] during utilization of the system.

Prototypical images are a good starting point for the development of an automated image classifier [6]. This knowledge is often collected by human experts in image catalogues. We describe based on a task for the study of the internal mitochondrial movement of cells [7] how such a classifier in combination with image analysis can be used for incremental knowledge acquisition and automatic classification. The work enhances our previous work on prototype-based classifier [2] by introducing the experts estimated similarity as new knowledge piece and a new function that adjusts this similarity and the automatically calculated similarity by the system in order to

P. Perner (Ed.): ICDM 2014, LNAI 8557, pp. 189–199, 2014.

improve the system accuracy. The test of the system is done on a new application on cell image analysis- the study of the internal mitochondrial movement of cells.

The classifier is set up based on prototypical cell appearances in the image such as for e.g. „healthy cell", „cell dead", and „cell in transition stage". For these prototypes are calculated image features based on random set theory that describes the texture on the cells. The prototype is represented then by the attribute-value pair and the class label. These settings are taken as initial classifier settings in order to acquire the knowledge about the dynamic signatures.

The importance of the features and the feature weights are learned by the protoclass-based classifier [2]. After the classifier is set up each new cell is then compared by the protoclass-based classifier and the similarity to the prototypes is calculated. If the similarity is high the new cell gets the label of the prototype. If the similarity to the prototypes is too low then there is evidence that the cell is in transition stage and a new prototype has been found. With this procedure we can learn the dynamic signature of the mitochondrial movement.

In Section 2 we present the methods for our prototype-based classifier. The material is described in Section 3 for the internal mitochondrial movement of cells. In Section 4 is presented the methodology for the knowledge acquisition based on a prototype-based classification. Results are given in Section 5 and finally in Section 6 conclusions are presented.

2   ProtoClass Classifiers

A prototype-based classifier classifies a new sample according to the prototypes in data base and selects the most similar prototype as output of the classifier. A proper similarity measure is necessary to perform this task but in most applications there is no a-priori knowledge available that suggests the right similarity measure. The method of choice to select the proper similarity measure is therefore to apply a subset of the numerous similarity measures known from statistics to the problem and to select the one that performs best according to a quality measure such as, for example, the classification accuracy. The other choice is to automatically build the similarity metric by learning the right attributes and attribute weights. The later one we chose as one option to improve the performance of our classifier.

When people collect prototypes to construct a dataset for a prototype-based classifier it is useful to check if these prototypes are good prototypes. Therefore a function is needed to perform prototype selection and to reduce the number of prototypes used for classification. This results in better generalization and a more noise tolerant classifier. If an expert selects the prototypes, this can result in bias and possible duplicates of prototypes causing inefficiencies. Therefore a function to assess a collection of prototypes and identify redundancy is useful.

Finally, an important variable in a prototype-based classifier is the value used to determine the number of closest cases and the final class label.

Consequently, the design-options the classifier has to improve its performance are prototype selection, feature-subset selection, feature weight learning and the 'k' value of the closest cases (see Figure 1).

We assume that the classifier can start in the worst case with only one prototype per class. By applying the classifier to new samples the system collects new prototypes. During the lifetime of the system it will chance its performance from an oracle-based classifier, which will classify the samples roughly into the expected classes, to a system with high performance in terms of accuracy.

In order to achieve this goal we need methods that can work on less number of prototypes and on large number of prototypes. As long as we have only a few numbers of prototypes feature subset selection and learning the similarity might be the important features the system needs. If we have more prototypes we also need prototype selection.

For the case with less number of prototypes we chose methods for feature subset selection based on the discrimination power of attributes. We use the feature based calculated similarity and the pair-wise similarity rating of the expert and apply the adjustment theory [Nie08] to fit the similarity value more to the true value.

For large number of prototypes we choose a decremental redundancy-reduction algorithm proposed by Chang [8] that deletes prototypes as long as the classification accuracy does not decrease. The feature-subset selection is based on the wrapper approach [9] and an empirical feature-weight learning method [10] is used. Cross validation is used to estimate the classification accuracy. A detailed description of our prototype-based classifier ProtoClass is given in [2]. The prototype selection, the feature selection, and the feature weighting steps are performed independently or in combination with each other in order to assess the influence these functions have on the performance of the classifier. The steps are performed during each run of the cross-validation process.

The classifier schema shown in Figure 1 is divided in the design phase (Learning Unit) and the normal classification phase (Classification Unit). The classification phase starts after we have evaluated the classifier and determined the right features, feature weights, the value for 'k' and the cases.

Our classifier has a flat data base instead of a hierarchical that makes it easier to conduct the evaluations.

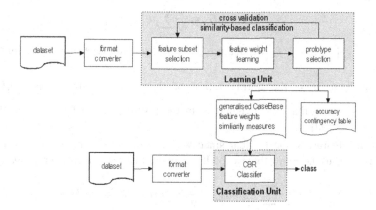

Fig. 1. Prototype-based Classifier

2.1   Classification Rule

Assume we have n prototypes that represent m classes of the application. Then, each new sample is classified based on its closeness to the n prototypes. The new sample is associated with the class label of the prototype that is the closest one to sample.

More precisely, we call $x'_n \in \{x_1,x_2,\ldots,x_i,\ldots x_n\}$ a closest case to x if $\min d(x_i, x) = d(x'_n, x)$, where $i=1,2,\ldots,n$.

The rule chooses to classify x into category C_l, where x'_n is the closest case to x and x'_n belongs to

class C_l with $l \in \{1,\ldots,m\}$.

In the case of the k-closest cases we require k samples of the same class to fulfill the decision rule. As a distance measure we can use any distance metric. In this work we used the city-block metric.

The pair-wise similarity measure Simij among our prototypes shows us the discrimination power of the chosen prototypes based on the features.

The calculated feature set must not be the optimal feature subset. The discrimination power of the features must be checked later. For a less number of prototypes we can let the expert judge the similarity SimEij $i, j \in \{1,\ldots,n\}$ between the prototypes. This gives us further information about the problem which can be used to tune the designed classifier.

2.2   Using Expert's Judgment on Similarity and the Calculated Similarity to Adjust the System

Humans can judge the similarity $SimE_{ij}$ among objects on a rate between 0 (identity) and 1(dissimilar). We can use this information to adjust the system to the true system parameters [11].

Using the city-block distance as distance measure, we get the following linear system of equations:

$$SimE_{ij} = \frac{1}{N} \sum_{l=1}^{N} a_l |f_{il} - f_{jl}| \qquad (1)$$

with $i, j \in \{1,\ldots,n\}$, f_{il} the feature l of the i-th prototype and N the number of attributes.

The attribute a_l is the normalization of the feature to the range $\{0,1\}$ with

$a_l = \dfrac{1}{|f_{max,l} - f_{min,l}|}$ that is calculated from the prototypes. That this is not the true

range of the feature value is clear since we have too less samples. The factor a_l is adjusted closer to the true value by the least square method using expert's $SimE_{ij}$:

$$\sum_{i=1}^{n-1} \sum_{j=i+1}^{n} \left(SimE_{ij} - \frac{1}{N} \sum_{l=1}^{N} a_l |f_{il} - f_{jl}| \right)^2 \Rightarrow Min! \qquad (2)$$

with the restriction $0 \leq a_l \leq \dfrac{1}{\left| f_{\max,l} - f_{\min l} \right|}$.

3   Methodology

Figure 2 summarizes the knowledge acquisition process based on protoclass-based classification.

We start with one prototype for each class. This prototype is chosen by the biologist based on the appearance of the cells. It requires that the biologist has enough knowledge about the processes going on in cell-based assays and can decide what kind of reaction the cell is showing.

The discrimination power of the prototypes is checked first based on the attributes values measured from the cells and the chosen similarity measure. Note that we calculated a large number of attributes for each cell. However many attributes does not mean that we will achieve a good discrimination power between the classes. It is better to come up with one or two attributes for small sample sizes in order to ensure a good performance of the classifier. The expert manually estimates the similarity between the prototypes and inputs these values into the system. The result of this process is the selection of the right similarity measure and the right number of attributes. With this information is set-up a first classifier and applied to real data.

Each new data gets associated with the label of the classification. Manually we evaluate the performance of the classifier. The biologist gives the true or gold label for the sample seen so far. This is kept into a data base and serves as gold standard for further evaluation. During this process the expert will sort out wrong classified data. This might happen because of too few prototypes for one class or because the samples should be divided into more classes. The decision what kind of technique should be applied is made based on the visual appearance of the cells. Therefore,

It is necessary to display the prototypes of class and the new samples. The biologist sorts these samples based on the visual appearance. That this is not easy to do by human is clear and needs some experiences in describing image information [6]. However, it is a standard technique in psychology in particular gestalts psychology known as categorizing or card sorting. As a result of this process we come up with more prototypes for one class or with new classes and at least one prototype for these new classes.

The discrimination power needs to get checked again based on this new data set. New attributes, new number of prototypes or a new similarity measure might be the output. The process is repeated as long as the expert is satisfied with the result. As result of the whole process we get a data set of samples with true class labels, the settings for the protoclass-based classifier, the important attributes and the real prototypes. The class labels represent the categories of the cellular processes going on in the experiment. The result can now be taken as a knowledge acquisition output. Just for discovering the categories or the classifier can now be used in routine work at the cell-line.

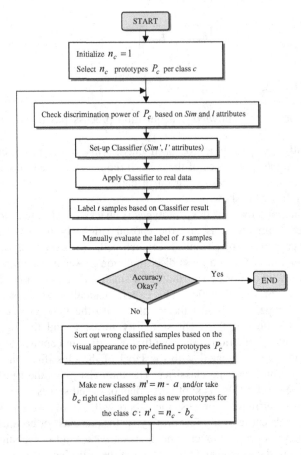

Fig. 2. Methodology for Prototype-based Classification

The discrimination power needs to get checked again based on this new data set. New attributes, new number of prototypes or a new similarity measure might be the output. The process is repeated as long as the expert is satisfied with the result. As result of the whole process we get a data set of samples with true class labels, the settings for the protoclass-based classifier, the important attributes and the real prototypes. The class labels represent the categories of the cellular processes going on in the experiment. The result can now be taken as a knowledge acquisition output. Just for discovering the categories or the classifier can now be used in routine work at the cell-line.

4   The Application

After the assay has been set up it is not quite clear what are the appearances of the different phases of a cell. This has to be learnt during the usage of the system.

Based on their knowledge the biologists set up several descriptions for the classification of the mitochondria. They grouped these classes in the following classes: tubular cells, round cells and dead cells. For the appearance of these classes see images in Figure 3.

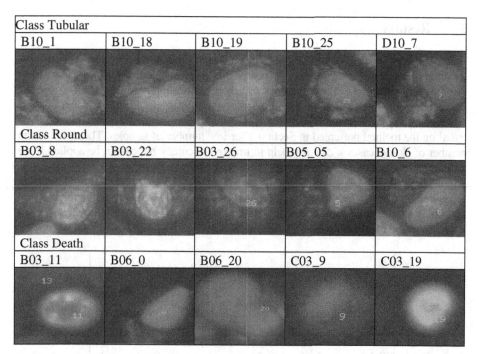

Class Tubular				
B10_1	B10_18	B10_19	B10_25	D10_7
Class Round				
B03_8	B03_22	B03_26	B05_05	B10_6
Class Death				
B03_11	B06_0	B06_20	C03_9	C03_19

Fig. 3. Sample Images for three Classes (top ClassTubular, middle Class Round, bottom Class Death)

Then prototypical cells were selected and the features were calculated with the software tool *CellInterpret* [12]. The expert rated the similarity between these prototypical images.

Our data set consist of 223 instances with the following class partition: 36 instances of class *Death*, 120 instances of class *Round*, 47 instances of class *Tubular*, and 114 features for each instance.

The expert chose for each class a prototype shown in Figure 4. The test data set for classification has then 220 instances. For our experiments we also selected 5 prototypes pro class respectively 20 prototypes pro class. The associate test data sets do not contain the prototypes.

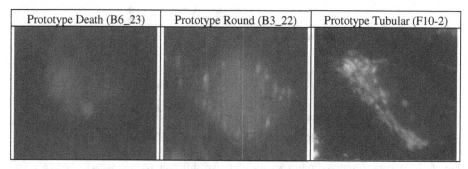

Prototype Death (B6_23)	Prototype Round (B3_22)	Prototype Tubular (F10-2)

Fig. 4. The Prototypes for the class Death, Round and Tubular

5   Results

Figure 5a shows the accuracy for classification based on different number of prototypes for all attributes and Fig. 5b shows the accuracy for a test set based on only the three most discriminating attributes. The test shows that the classification accuracy is not so bad for only three prototypes but with the number of prototypes the accuracy increases. The selection of the right subset of features can also improve the accuracy and can be done based on the method presented in Section 2 for low number of samples. The right chosen number of closest cases k can also help to improve accuracy but cannot be applied if we only have three prototypes or less prototypes in the data base.

Figure 6 shows the classification results for the 220 instances started without adjustment meaning the weights al are equal to one (1;1;1) and with adjustment based on expert`s rating where the weights are (0.00546448; 0.00502579;0.00202621) as an outcome of the minimization problem.

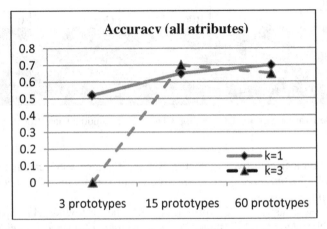

Fig. 5. a. Accuracy versus Prototypes and for two different feature subset; Accuracy for different number of prototypes using all attributes

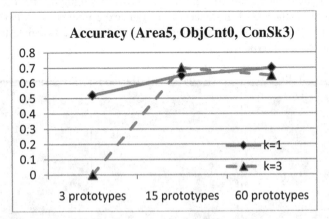

Fig. 6. b. Accuracy versus Prototypes and for two different feature subset; Accuracy for different number of prototypes using 3 attributes (Area5, ObjCtn0, ConSk3)

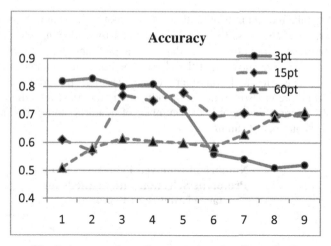

Fig. 7. Accuracy depending on choice of attributes (k=1)

Table 1. Difference between 3 Prototypes using the 3 attributes (ObjCnt0,ArSig0, ObjCnt1)

	B6_23	B03_22	F10_2
B6_23	0	0,669503257 (0,8)	0,989071038 (0,6)
B03_22	0,669503257 (0,8)	0	0,341425705 (0,9)
F10_2	0,989071038 (0,6)	0,341425705 (0,9)	0

Table 1 shows the difference values of three prototypes. The result shows that accuracy can be improved by applying the adjustment theory and especially the class specific quality is improved by applying the adjustment theory (see Fig. 7).

The application of the methods for larger samples set did not bring any significant reduction in the number of prototypes (see Fig. 8) or in the feature subset (see Fig. 9). The prototype selection method reduced the number of prototypes only by three

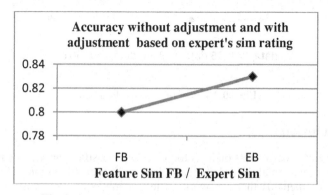

Fig. 8. Accuracy with and without adjustment theory

prototypes. We take it as an indication that we have not yet the enough prototypes and that the accuracy of the classifier can be improved by collecting more prototypes. How these functions worked on another data set can be found in [LCSP08b].

In Summary, we have shown that the chosen methods are valuable methods for a prototype-based classifier and can improve the classifier performance. For future work we will do more investigations on the adjustment theory as a method to learn the importance of features based on less number of features and for feature subset selection for less number of samples.

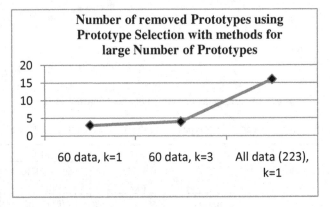

Fig. 9. Number of removed Prototypes

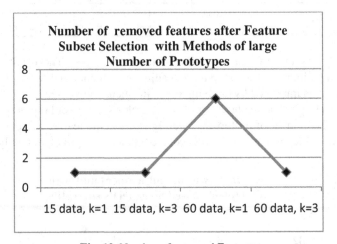

Fig. 10. Number of removed Features

6   Conclusions

We have presented our results on a prototype-based classification. Such a method can be used for incremental knowledge acquisition and classification. Therefore the classifier needs methods that can work on less numbers of prototypes and on large number of prototypes. Our result shows that feature subset selection based on the discrimination power of a feature is a good method for less numbers of prototypes. The adjustment theory in combination with an expert similarity judgment can be taken to learn the true feature range in case of less prototypes. If we have large number of prototypes an option for prototype selection that can check for redundant prototypes is necessary.

The system can start to work on a low number of prototypes and can instantly collect samples during the usage of the system. These samples get the label of the closest case. The system performance improves as more prototypes the system has in its data base. That means an iterative process of labeled sample collection based on prototype based classification is necessary followed by a revision of these samples after some time in order to sort out wrong classified samples until the system performance has been stabilized.

The test of the system is done on a new application on cell image analysis, the study of the internal mitochondrial movement of cells.

Acknowledgement. This work has been sponsored by the German Ministry of Science and Technology BMBF under the grant "Quantitative Measurement of Dynamic Time Dependent Cellular Events, QuantPro" grant no. 0313831B.

References

1. Schmidt, R., Gierl, L.: Temporal Abstractions and Case-Based Reasoning for Medical Course Data: Two Prognostic Applications. In: Perner, P. (ed.) MLDM 2001. LNCS (LNAI), vol. 2123, pp. 23–34. Springer, Heidelberg (2001)
2. Perner, P.: Prototype-Based Classification. Applied Intelligence 28, 238–246 (2008)
3. Nilsson, M., Funk, P.: A Case-Based Classification of Respiratory Sinus Arrhythmia. In: Funk, P., González Calero, P.A. (eds.) ECCBR 2004. LNCS (LNAI), vol. 3155, pp. 673–685. Springer, Heidelberg (2004)
4. Aha, D.W., Kibler, D., Albert, M.K.: Instance-based Learning Algorithm. Machine Learning 6(1), 37–66 (1991)
5. Bichindaritz, I., Kansu, E., Sullivan, K.M.: Case-Based Reasoning in CARE-PARTNER: Gathering Evidence for Evidence-Based Medical Practice. In: Smyth, B., Cunningham, P. (eds.) EWCBR 1998. LNCS (LNAI), vol. 1488, pp. 334–345. Springer, Heidelberg (1998)
6. Sachs-Hombach, K.: Bildbegriff und Bildwissenschaft. In: Gerhardus, D., Rompza, S. (eds.) Kunst - Gestaltung - Design, Heft 8, pp. 1–38. Verlag St. Johann, Saarbrücken (2002)
7. Krausz, E., Prechtl, S., Stelzer, E.H.K., Bork, P., Perner, P.: Quantitative Measurement of dynamic time dependent cellular events. Project Description (May 2006)
8. Chang, C.-L.: Finding Prototypes for Nearest Neighbor Classifiers. IEEE Trans. on Computers C-23(11) (1974)
9. Perner, P. (ed.): Data Mining on Multimedia Data. LNCS, vol. 2558. Springer, Heidelberg (2002)
10. Little, S., Colantonio, S., Salvetti, O., Perner, P.: Evaluation of Feature Subset Selection, Feature Weighting, and Prototype Selection for Biomedical Applications. J. Software Engineering & Applications 3, 39–49 (2010)
11. Niemeier, W.: Ausgleichsrechnung. de Gruyter, Berlin (2008)
12. Perner, P.: Novel Computerized Methods in System Biology–Flexible High-Content Image Analysis and Interpretation System for Cell Images. In: Perner, P., Salvetti, O. (eds.) MDA 2008. LNCS (LNAI), vol. 5108, pp. 139–157. Springer, Heidelberg (2008)

Discovering Main Vertexical Planes in a Multivariate Data Space by Using CPL Functions

Leon Bobrowski

Faculty of Computer Science, Bialystok University of Technology, Bialystok, Poland
l.bobrowski@pb.edu.pl

Abstract. Data mining problems and tools are linked to the task of extracting important regularities (patterns) from multivariate data sets. In some cases, flat patterns can be located on vertexical planes in a multidimensional data space. Vertexical planes are linked to vertices in parameter space. Patterns located on vertexical planes can be discovered in large data sets through minimization of the convex and piecewise linear (*CPL*) criterion functions.

Keywords: high dimensional data sets, data mining, *CPL* criterion functions, main vertexical planes, flat patterns, *K-planes* clustering.

1   Introduction

Data mining tasks are aimed at discovering useful patterns in large data sets [1], [2], [3]. The term *patterns* stands for various types of regularities in an explored data set, such as decision rules, trends, or models of interactions. The extracted patterns are used in solving many practical problems linked, e.g. to medical diagnosis support, economic forecasting, marketing, fraud detection or to scientific discoveries.

Data sets can typically be represented as clouds of points in a multidimensional feature space. Clustering algorithms belong to the most powerful tools of data mining [2]. The *K-means* algorithms constitute the most popular and successful paradigm in clustering applications. The basic idea in the *K-means* algorithm is linked to partitioning of a given set C of m objects (points) into K subsets C_k centered around the class prototypes in the form of *central points*, which had been computed (defined) earlier. In the next step, the class prototypes are modified in accordance with the obtained subsets C_k. These steps are repeated until the central points are stabilized in the successive steps. The *K-means* algorithm has been generalized to the form of *K-means*, *K-planes* or *K-models* [4], [5]. In these approaches the concept of *central points* has been replaced by *central models*, e.g. in the form of planes in multivariate feature space.

Central planes can have the form of *vertexical planes* based on vertices in parameter space. *"Flat patterns"* located on the *main vertexical planes* can be discovered in large data sets through minimization of the convex and piecewise linear (*CPL*) criterion functions [6]. Analytical and computational properties of the method of main vertexical planes discovering through minimization of the *CPL* criterion functions are analyzed in the presented article.

P. Perner (Ed.): ICDM 2014, LNAI 8557, pp. 200–213, 2014.
© Springer International Publishing Switzerland 2014

2   Hyperplanes and Vertices in the Parameter Space

Let us assume that the data set C contains m feature vectors $\mathbf{x}_j[n] = [x_{j1},...,x_{jn}]^T$ belonging to a given n-dimensional feature space $F[n]$ ($\mathbf{x}_j[n] \in F[n]$):

$$C = \{\mathbf{x}_j[n]\}, \quad where \quad j = 1,...,m \tag{1}$$

Components x_{ji} of the feature vector $\mathbf{x}_j[n]$ can be treated as the numerical results of n standardized examinations of a given object O_j ($x_{ji} \in \{0,1\}$ or $x_{ji} \in R$).

We can assume without limitation that the feature space $F[n]$ is equal to the n-dimensional space of real numbers R^n ($F[n] = R^n$). Each feature vector $\mathbf{x}_j[n]$ can be treated as a point in the space R^n.

Each of m feature vector $\mathbf{x}_j[n]$ from the set C (1) defines the below hyperplane h_j in the parameter space R^n ($\mathbf{w}[n] \in R^n$):

$$(\forall \mathbf{x}_j[n] \in C) \qquad h_j = \{\mathbf{w}[n]: \mathbf{x}_j[n]^T\mathbf{w}[n] = 1\} \tag{2}$$

Each unit vector $\mathbf{e}_i[n] = [0,...,1,...,0]^T$ defines the below hyperplane h_i^0 in the parameter space R^n:

$$(\forall i \in \{1,...,n\}) \qquad h_i^0 = \{\mathbf{w}[n]: \mathbf{e}_i[n]^T\mathbf{w}[n] = 0\} = \\ = \{\mathbf{w}[n]: w_i = 0\} \tag{3}$$

Let us consider a set S_k of n linearly independent feature vectors $\mathbf{x}_j[n]$ ($j \in J_k$) and unit vectors $\mathbf{e}_i[n]$ ($i \in I_k$).

$$S_k = \{\mathbf{x}_j[n]: j \in J_k\} \cup \{\mathbf{e}_j[n]: i \in I_k\} \tag{4}$$

The k-th *vertex* $\mathbf{w}_k[n]$ in the parameter space R^n is the intersection point of n hyperplanes h_i (2) or h_i^0 (3) defined by the vectors $\mathbf{x}_j[n]$ ($j \in J_k$) and $\mathbf{e}_i[n]$ ($i \in I_k$) from the set S_k (4). The intersection point $\mathbf{w}_k[n]$ can be given by the below linear equations:

$$(\forall j \in J_k) \qquad \mathbf{w}_k[n]^T\mathbf{x}_j[n] = 1 \tag{5}$$

and

$$(\forall i \in I_k) \qquad \mathbf{w}_k[n]^T\mathbf{e}_i[n] = 0 \tag{6}$$

The equations (5) and (6) can be given in the below matrix form:

$$\mathbf{B}_k[n]\,\mathbf{w}_k[n] = \mathbf{1}'[n] = [1,...,1,0,...,0]^T \tag{7}$$

where $\mathbf{B}_k[n]$ is the square matrix, the k-th *basis* linked to the vertex $\mathbf{w}_k[n]$:

$$\mathbf{B}_k[n] = [\mathbf{x}_{j(1)}[n],...,\mathbf{x}_{j(n')}[n],\mathbf{e}_{i(n'+1)}[n],...,\mathbf{e}_{i(n)}[n]]^T \tag{8}$$

and

$$\mathbf{w}_k[n] = \mathbf{B}_k[n]^{-1}\mathbf{1}'[n] \tag{9}$$

The number of the subsets S_k (4) and the bases $\mathbf{B}_k[n]$ (8) could be very large. The same vertex $\mathbf{w}_k[n]$ can be determined by (9) more than one base $\mathbf{B}_k[n]$.

Definition 1: The *rank* r_k $(1 \le r_k \le n)$ of the vertex $\mathbf{w}_k[n]$ (9) is defined as the number of the feature vectors $\mathbf{x}_j[n]$ $(j \in J_k$ (5)) in the base $\mathbf{B}_k[n]$ (8) linked (9) to this vertex.

It can be noted that the *rank* r_k of the vertex $\mathbf{w}_k[n] = [w_{k,1},\ldots,w_{k,n}]^T$ (9) is equal to the number of its nonzero components $w_{k,i}$ $(w_{k,i} \ne 0)$.

Definition 2: The vertex $\mathbf{w}_k[n]$ of the rank r_k is degenerated when more than r_k hyperplanes h_j (2) pass (5) through it.

Let us note that the degenerated vertex $\mathbf{w}_k[n]$ can be defined (9) by at least two different matrices $\mathbf{B}_k[n]$ and $\mathbf{B}_{k'}[n]$ $(\mathbf{B}_k[n] \ne \mathbf{B}_{k'}[n]$ and $\mathbf{w}_k[n] = \mathbf{w}_{k'}[n])$.

Definition 3: The *degree of degeneration* of the vertex $\mathbf{w}_k[n]$ (9) of the rank r_k is defined as the number $d_k = m_k - r_k$, where m_k is the number of such feature vectors $\mathbf{x}_j[n]$ $(\mathbf{x}_j[n] \in C)$ from the set C (1), which define the hyperplanes h_j (2) passing through this vertex $(\mathbf{w}_k[n]^T\mathbf{x}_j[n] = 1)$.

The *degree of degeneration* d_k of the vertex $\mathbf{w}_k[n]$ (9) can be also seen as the number of different bases $\mathbf{B}_{k'}[n]$ (8) linked (9) to this vertex. The *degree of degeneration* of the vertex $\mathbf{w}_k[n]$ (9) can be defined also in a different way, for example as $d_k' = (m_k - r_k) / (m - r_k)$. It could be seen that $0 \le d_k' \le 1$.

3   Hyperplanes and Planes in the Feature Space

The hyperplanes $H(\mathbf{w}[n],\theta)$ in the feature space $F[n]$ are usually defined in the below manner [2]:

$$H(\mathbf{w}[n], \theta) = \{\mathbf{x}[n]: \mathbf{w}[n]^T\mathbf{x}[n] = \theta\} \tag{10}$$

where $\mathbf{w}[n] = [w_1,\ldots,w_n]^T$ is the *weight vector* $(\mathbf{w}[n] \in R^n)$ and θ is the *threshold* $(\theta \in R^1)$.

The *weight vector* $\mathbf{w}[n]$ is perpendicular to $H(\mathbf{w}[n],\theta)$ and determines the orientation of this hyperplane. Changing the threshold θ causes a parallel displacement (shift) of the hyperplane $H(\mathbf{w}[n],\theta)$ (2). The dimension of the hyperplane $H(\mathbf{w}[n],\theta)$ (10) is equal to $n - 1$.

The vertex $\mathbf{w}_k[n]$ (9) of the rank r_k allows to define the $(r_k - 1)$ - dimensional *vertexical plane* $P_k(\mathbf{x}_{j(1)}[n],\ldots, \mathbf{x}_{j(rk)}[n])$ in the feature space $F[n]$ as the linear combination of r_k (5) feature vectors $\mathbf{x}_{j(i)}[n]$ belonging to the basis $\mathbf{B}_k[n]$ (8):

$$P_k(\mathbf{x}_{j(1)}[n],\ldots,\mathbf{x}_{j(rk)}[n]) = \{\mathbf{x}[n]: \mathbf{x}[n] = \alpha_1 \mathbf{x}_{j(1)}[n] +\ldots+ \alpha_k \mathbf{x}_{j(rk)}[n]\} \tag{11}$$

where $j(i) \in J_k$ (5) and r_k parameters α_i $(\alpha_i \in R^1)$ fulfill the below condition:

$$\alpha_1 +\ldots+ \alpha_{rk} = 1 \tag{12}$$

If the vertex $\mathbf{w}_k[n]$ (9) has the rank $r_k = n$, then the vertexical plane (11) has the dimension equal to $(n - 1)$, similarly to the hyperplane $H(\mathbf{w}[n],\theta)$ (2). We can note that not every hyperplane $H(\mathbf{w}[n],\theta)$ (2) can be represented as $P_k(\mathbf{x}_{j(1)}[n],\ldots, \mathbf{x}_{j(n)}[n])$ (11)

but the opposite statement is true. Every vertexical hyperplane $P_k(\mathbf{x}_{j(1)}[n],\ldots, \mathbf{x}_{j(n)}[n])$ (11) can be represented as $H(\mathbf{w}[n],\theta)$ (2).

Remark 1: The formula (11) without the condition (12) defines such an r_k - dimensional plane $P_k(\mathbf{x}_{j(1)}[n],\ldots, \mathbf{x}_{j(rk)}[n])$ in the feature space $F[n]$ which passes through the point zero $\mathbf{0}[n]$ (*origin*).

The *line* $L_k(\mathbf{x}_{j(1)}[n],\mathbf{x}_{i(2)}[n])$ in the feature space $F[n]$ can be treated as the one-dimensional *plane* $P_k(\mathbf{x}_{i(1)}[n], \mathbf{x}_{i(2)}[n])$ (11) spanned by two different vectors $\mathbf{x}_{j(1)}[n]$ and $\mathbf{x}_{j(2)}[n]$ $(\mathbf{x}_{j(1)}[n] \neq \mathbf{x}_{j(2)}[n])$ by using only one parameter α $(\alpha \in R^1)$:

$$L_k(\mathbf{x}_{j(1)}[n],\mathbf{x}_{j(2)}[n]) = \{\mathbf{x}[n]: \mathbf{x}[n] = \mathbf{x}_{j(1)}[n] + \alpha\,(\mathbf{x}_{j(2)}[n] - \mathbf{x}_{j(1)}[n])\} = \qquad (13)$$
$$= \{\mathbf{x}[n]: \mathbf{x}[n] = (1 - \alpha)\,\mathbf{x}_{j(1)}[n] + \alpha\,\mathbf{x}_{j(2)}[n]\}$$

One feature vector $\mathbf{x}_j[n]$ allows to define the line $L_0(\mathbf{x}[n])$ passing through the point $\mathbf{0}[n]$ (origin) of the feature space $F[n]$:

$$L_0(\mathbf{x}_{j(1)}[n]) = \{\mathbf{x}[n]: \mathbf{x}[n] = \alpha\,\mathbf{x}_{j(1)}[n]\}, \text{ where } \alpha \in R^1 \qquad (14)$$

Remark 2: If two feature vectors $\mathbf{x}_{i(1)}[n]$ and $\mathbf{x}_{i(2)}[n]$ are linearly dependent $(\mathbf{x}_{i(2)}[n] = c\,\mathbf{x}_{i(1)}[n], \text{ where } c \in R^1)$, then the line $L(\mathbf{x}_{i(1)}[n],\mathbf{x}_{i(2)}[n])$ (13) passes through the origin, and the equation (13) can be reduced to (14).

Theorem 1: The feature vector $\mathbf{x}_i[n]$ defines hyperplane h_i (2) which passes through the vertex $\mathbf{w}_k[n]$ (9) of the rank r_k if and only if the vector $\mathbf{x}_i[n]$ is situated on the (r_k-1)-dimensional *vertexical plane* $P_k(\mathbf{x}_{j(1)}[n],\ldots,\mathbf{x}_{j(rk)}[n])$ (11), where $j(i) \in J_k$ (5).

Proof: Each vector $\mathbf{x}_{i(i)}[n]$ belonging to the basis $\mathbf{B}_k[n]$ (8) fulfils the equation $\mathbf{w}_k[n]^T\mathbf{x}_{i(i)}[n] = 1$ (5). If the point $\mathbf{x}_i[n]$ is situated on the $(r_k - 1)$-dimensional vertexical plane $P_k(\mathbf{x}_{i(1)}[n],\ldots, \mathbf{x}_{i(rk)}[n])$ (11) then it satisfies the equation $\mathbf{x}_j[n] = \alpha_1\mathbf{x}_{i(1)}[n] + \ldots + \alpha_k\mathbf{x}_{j(rk)}[n]$, where $\alpha_1 + \ldots + \alpha_{rk} = 1$ (6). Therefore, the condition $\mathbf{w}_k[n]^T\mathbf{x}_j[n] = 1$ results.

Any feature vector $\mathbf{x}_j[n]$ can be represented as the linear combination of the basis vectors $\mathbf{x}_{j(i)}[n]$ $(j(i) \in J_k)$ (5) and $\mathbf{e}_j[n]$ $(i \in I_k)$ (6). The basis unit vectors $\mathbf{e}_j[n]$ fulfill the condition $\mathbf{w}_k[n]^T\mathbf{e}_j[n] = 0$ (6). Therefore, if the equation $\mathbf{w}_k[n]^T\mathbf{x}_j[n] = 1$ holds for some vector $\mathbf{x}_j[n]$, then the condition (6) implies that this vector is situated on the vertexical plane $P_k(\mathbf{x}_{j(1)}[n],\ldots,\mathbf{x}_{j(rk)}[n])$ (11) in the feature space $F[n]$.

We are interested in discovering such a vertexical plane $P_k(\mathbf{x}_{i(1)}[n],\ldots, \mathbf{x}_{j(rk)}[n])$ (11) which would contain especially numerous feature vectors $\mathbf{x}_j[n]$ (1).

Definition 4: The *vertexical plane* $P_k(\mathbf{x}_{i(1)}[n],\ldots, \mathbf{x}_{j(rk)}[n])$ (11) that includes m_k feature vectors $\mathbf{x}_j[n]$ from the data set C (1) is called the *main vertexical plane* if and only if the number m_k is large in comparison to the *rank* r_k (*Definition* 2) of the vertex $\mathbf{w}_k[n]$ (at least $m_k > r_k$).

Supposition I: Discovering *main vertexical planes* $P_k(\mathbf{x}_{i(1)}[n],\ldots, \mathbf{x}_{j(rk)}[n])$ (11) in the feature space $F[n]$ can be based on the detection and examination of the vertices $\mathbf{w}_k[n]$ (9) with a high *degree of degeneration* r_k (*Definition* 3).

4   Convex and Piecewise Linear (*CPL*) Criterion Functions

Let us consider a convex and piecewise linear (*CPL*) penalty functions $\varphi_j(\mathbf{w})$ defined in the below manner on the feature vectors $\mathbf{x}_j[n]$ from the data set C (1) [6]:

$(\forall \mathbf{x}_j[n] \in C\,(1))$

$$\varphi_j(\mathbf{w}[n]) = \begin{cases} 1 - \mathbf{w}[n]^T\mathbf{x}_j[n] & \textit{if} \quad \mathbf{w}[n]^T\mathbf{x}_j[n] \le 1 \\ \mathbf{w}[n]^T\mathbf{x}_j[n] - 1 & \textit{if} \quad \mathbf{w}[n]^T\mathbf{x}_j[n] > 1 \end{cases} \qquad (15)$$

The penalty functions $\varphi_j(\mathbf{w})$ are equal to the absolute values $|1 - \mathbf{w}[n]^T\mathbf{x}_j[n]|$.

The criterion function $\Phi_m(\mathbf{w}[n])$ is defined as the weighted sum of the penalty functions $\varphi_j(\mathbf{w}[n])$ defined by feature vectors $\mathbf{x}_j[n]$ from the subset C_m $(C_m \subset C)$:

$$\Phi_m(\mathbf{w}[n]) = \sum_{i \in Jm} \alpha_j\, \varphi_j(\mathbf{w}[n]) \qquad (16)$$

where $J_m = \{j: \mathbf{x}_j[n] \in C_m\}$ and the positive parameters α_j ($\alpha_j > 0$) in the below function $\Phi_m(\mathbf{w}[n])$ can be treated as the *prices* of particular feature vectors $\mathbf{x}_j[n]$. The standard choice of the parameters α_j values is one:

$$(\forall j \in J_m)\ \ \alpha_j = 1.0 \qquad (17)$$

The criterion function $\Phi_m(\mathbf{w}[n])$ (16) is convex and piecewise linear as the sums of the *CPL* functions $\alpha_j\varphi_j(\mathbf{w}[n])$ (15). It can be proved that the minimal value of the function $\Phi_m(\mathbf{w}[n])$ can be found in one of the vertices $\mathbf{w}_k[n]$ (9) [7]:

$$(\exists \mathbf{w}_k^*[n])\ \ (\forall \mathbf{w}[n])\ \ \Phi_m(\mathbf{w}[n]) \ge \Phi_m(\mathbf{w}_k^*[n]) = \Phi_m^* \ge 0 \qquad (18)$$

The basis exchange algorithms which are similar to the linear programming allow to find efficiently the minimum $\Phi_m(\mathbf{w}_k^*[n])$ of the criterion functions $\Phi_m(\mathbf{w}[n])$ (16) even in the case of large, multidimensional data subsets C_m $(C_m \subset C)$ (1) [8].

Theorem 2: The minimal value $\Phi_m(\mathbf{w}_k^*[n])$ (18) of the criterion function $\Phi_m(\mathbf{w}[n])$ (16) is equal to zero $(\Phi_m(\mathbf{w}_k^*[n]) = 0)$, if and only if all the feature vectors $\mathbf{x}_j[n]$ from the subset C_m $(C_m \subset C\,(1))$ are situated on a hyperplane $H(\mathbf{w}[n],\theta)$ (10) with $\theta \ne 0$.

Proof: Let us suppose that all the feature vectors $\mathbf{x}_j[n]$ from the subset C_m are situated on some hyperplane $H(\mathbf{w}'[n],\theta')$ (10) with $\theta' \ne 0$:

$$(\forall \mathbf{x}_j[n] \in C_m)\ \ \mathbf{w}'[n]^T\mathbf{x}_j[n] = \theta' \qquad (19)$$

From this

$$(\forall \mathbf{x}_j[n] \in C_m)\ \ (\mathbf{w}'[n]\,/\,\theta')^T\mathbf{x}_j[n] = 1 \qquad (20)$$

The above equations mean that functions $\varphi_j(\mathbf{w}'[n]\,/\,\theta')$ (15) are equal to zero in the point $(\mathbf{w}'[n]\,/\,\theta')$:

$$(\forall \mathbf{x}_j[n] \in C_m)\ \ \varphi_j(\mathbf{w}'[n]\,/\,\theta') = 0 \qquad (21)$$

so

$$\Phi_m(\mathbf{w}'[n] / \theta') = 0 \tag{22}$$

On the other hand, if the criterion function $\Phi_m(\mathbf{w}'[n])$ (16) is equal to zero in some point $\mathbf{w}'[n]$, then each of the penalty functions $\varphi_j(\mathbf{w}'[n])$ (15) has to be equal to zero:

$$(\forall \mathbf{x}_j[n] \in C_m) \quad \varphi_j(\mathbf{w}'[n]) = 0 \tag{23}$$

or

$$(\forall \mathbf{x}_j[n] \in C_m) \quad \mathbf{w}'[n]^T \mathbf{x}_j[n] = 1 \tag{24}$$

The above equations mean that each feature vector $\mathbf{x}_j[n]$ from the subset C_m is located on the hyperplane $H(\mathbf{w}'[n], 1)$ (10). $\Updownarrow$

Taking into account that the minimal value (18) of the criterion function $\Phi_m(\mathbf{w}[n])$ (16) can be located in one of the vertices $\mathbf{w}_k[n]$ (9), the *Theorem* 2 can be reformulated in the below manner:

Theorem 2': The minimal value $\Phi_m(\mathbf{w}_k^*[n])$ (18) of the criterion function $\Phi_m(\mathbf{w}[n])$ (16) is equal to zero ($\Phi_m(\mathbf{w}_k^*[n]) = 0$), if and only if all the feature vectors $\mathbf{x}_i[n]$ from the subset C_m ($C_m \subset C$ (1)) are situated on some vertexical plane $P_k(\mathbf{x}_{j(1)}[n],..., \mathbf{x}_{j(l)}[n])$ (11) (12) which does not pass through the point zero $\mathbf{0}[n]$ (*origin*).

The below theorem characterizes the *invariance property* of the value $\Phi_k(\mathbf{w}_k^*[n])$:

Theorem 3: The minimal value $\Phi_m(\mathbf{w}_k^*[n])$ (18) of the criterion function $\Phi_m(\mathbf{w}[n])$ (16) does not depend on linear, non-singular data transformations of the feature vectors $\mathbf{x}_j[n]$ from the subset C_m ($C_m \subset C$ (1)):

$$\Phi_m'(\mathbf{w}_k'[n]) = \Phi_m(\mathbf{w}_k^*[n]) \tag{25}$$

where $\Phi_m'(\mathbf{w}_k'[n])$ is the minimal value of the criterion functions $\Phi_m'(\mathbf{w}[n])$ (16) defined on the transformed feature vectors $\mathbf{x}_j'[n]$:

$$(\forall \mathbf{x}_j[n] \in C_m) \quad \mathbf{x}_j'[n] = A[n] \, \mathbf{x}_j[n] \tag{26}$$

where $A[n]$ is a non-singular matrix of dimension ($n \times n$) ($A^{-1}[n]$ exists).

Proof: The values $\varphi_i'(\mathbf{w}[n])$ of the penalty function $\varphi_i(\mathbf{w}[n])$ (15) in a point $\mathbf{w}'[n]$ are defined in the below manner on the transformed feature vectors $\mathbf{x}_j'[n]$ (26):

$$(\forall \mathbf{x}_j'[n] \in C_m) \quad \varphi_j'((\mathbf{w}'[n]) = |1 - \mathbf{w}'[n]^T \mathbf{x}_j'[n]| = |1 - \mathbf{w}'[n]^T A[n] \mathbf{x}_j[n]| \tag{27}$$

If we take

$$\mathbf{w}_k'[n] = (A[n]^T)^{-1} \mathbf{w}_k^*[n] \tag{28}$$

we obtain the below result

$$(\forall \mathbf{x}_j[n] \in C_m) \quad \varphi_j'(\mathbf{w}_k'[n]) = \varphi_j(\mathbf{w}_k^*[n]) \tag{29}$$

The above equations mean that the value $\Phi_m'(\mathbf{w}_k'[n])$ of the criterion functions $\Phi_m'(\mathbf{w}[n])$ (16) defined in the point $\mathbf{w}_k'[n]$ (28) on the transformed feature vectors

$x_i'[n]$ (26) is equal to the minimal value $\Phi_m(w_k^*[n])$ (18) of the criterion function $\Phi_m(w[n])$ (16) defined on the feature vectors $x_j[n]$.

The minimal value $\Phi_m(w_k^*[n])$ (18) of the criterion function $\Phi_m(w[n])$ (16) can be characterized by two below *monotonocity properties*:

i. The positive monotonocity property due to reduction of feature vectors $x_i[n]$
Neglecting some feature vectors $x_i[n]$ cannot result in an increase of the minimal value $\Phi_m(w_m^*[n])$ (19) of the criterion function $\Phi_m(w[n])$ (17):

$$(C_{m'} \subset C_m) \Rightarrow (\Phi_{m'}^* \leq \Phi_m^*) \tag{30}$$

where the symbol $\Phi_{m'}^*$ stands for the minimal value (18) of the criterion function $\Phi_{m'}(w[n])$ (16) defined on the elements $x_j[n]$ of the subset $C_{m'}$ ($x_j[n] \in C_{m'}$).

The relation (30) can be justified by the remark that neglecting some feature vectors $x_j[n]$ results in neglecting some non-negative components $\alpha_j \varphi_j(w[n])$ (15) in the criterion function $\Phi_m(w[n])$ (16).

ii. The negative monotonicity property due to reduction of features x_i
The reduction of the feature space $F[n]$ to $F'[n']$ by neglecting some features x_i cannot result in a decrease of the minimal value $\Phi_k(w_k^*[n])$ (18) of the criterion function $\Phi_k(w[n])$ (16):

$$(F'[n'] \subset F[n]) \Rightarrow (\Phi_{m'}^* \geq \Phi_m^*) \tag{31}$$

where the symbol $\Phi_{m'}^*$ stands for the minimal value (18) of the criterion function $\Phi_m(w[n'])$ (16) defined on the reduced vectors $x_i'[n']$ ($x_i'[n'] \in F'[n']$, $n' < n$). The relation (31) results from the fact that the neglecting some features x_i is equivalent to imposing additional constraints $"w_i = 0"$ in the parameter space R^n..

5   Procedure of the Main Vertexical Planes Discovering

The feature vector $x_i[n]$ is *included* in the vertexical plane $P(x_{j(1)}[n],...,x_{j(rk)}[n])$ (11) if the below equation holds:

$$x_j[n] = \alpha_{j,1} x_{j(1)}[n] +...+ \alpha_{j,rk} x_{j(rk)}[n] \tag{32}$$

with the condition (12):

$$\alpha_{j,1} +...+ \alpha_{j,rk} = 1 \tag{33}$$

Definition 4: The *vertexical plane* $P_k(x_{i(1)}[n],..., x_{j(rk)}[n])$ (11) that includes m_k feature vectors $x_j[n]$ from the data set C (1) is called the *main vertexical plane* if and only if the number m_k is a large in comparison with to the *rank r_k* (*Definition 2*) of the vertex $w_k[n]$ (at least $m_k > r_k$).

The below multistage *Procedure Vertex* is proposed for discovering the main vertical plane $P_m(\mathbf{x}_{j(1)}[n],...,\mathbf{x}_{j(rk)}[n])$ (11) on the basis of the data set $C_m = C$ (1):

$$\textit{Procedure Vertex} \qquad (34)$$

i.   Find the minimal value $\Phi_m(\mathbf{w}_k^*[n])$ (18) and the optimal vertex $\mathbf{w}_k^*[n]$ of the criterion function $\Phi_m(\mathbf{w}[n])$ (16) defined on elements $\mathbf{x}_j[n$ of the subset C_m

ii. If $\Phi_m(\mathbf{w}_k^*[n]) = 0$, then the *Procedure Vertex* is **stopped** in the optimal vertex $\mathbf{w}_k^*[n]$, otherwise the next step is executed

iii. Find the vector $\mathbf{x}_{j'}[n]$ in the feature subset C_m with the highest value of the penalty function $\varphi_j(\mathbf{w}[n])$ (15) in the optimal vertex $\mathbf{w}_k^*[n]$ (18)

$$(\forall \mathbf{x}_j[n] \in C_m) \quad \varphi_{j'}(\mathbf{w}_k^*[n]) \geq \varphi_j(\mathbf{w}_k^*[n]) \qquad (35)$$

or with the parameters α_j (16) taking into account:

$$(\forall \mathbf{x}_j[n] \in C_m) \quad \alpha_{j'}\varphi_{j'}(\mathbf{w}_k^*[n]) \geq \alpha_j \, \varphi_j(\mathbf{w}_k^*[n]) \qquad (36)$$

iv. Remove the feature vector $\mathbf{x}_{j'}[n]$ from the subset C_m $(C_m \rightarrow C_m / \{ \mathbf{x}_{j'}[n] \})$ and go to the step i.

It can be proved that the *Procedure Vertex* is **stopped** in the optimal vertex $\mathbf{w}_k^*[n]$ (18) of the rank r_k after finite number of steps. This property is based on the *Theorem 2* and on the *monotonocity property* (30).

Let the symbol C_m^* stand for optimal subset of feature vectors $\mathbf{x}_j[n]$ which is obtained when the *Procedure Vertex* is **stopped**.

Remark 3: The *degree of degeneration* d_k of the optimal vertex $\mathbf{w}_k^*[n]$ (18) (*Definition* 3) is equal to the difference between the number m_k of elements $\mathbf{x}_j[n]$ of the optimal subset C_m^* and the *rank* r_k of this vertex (*Definition* 2):

$$d_k = m_k - r_k \qquad (37)$$

In accordance with the *Definition* 4, the *main vertexical plane* $P_k(\mathbf{x}_{j(1)}[n],...,$ $\mathbf{x}_{j(rk)}[n])$ (11) based on the optimal vertex $\mathbf{w}_k^*[n]$ (18) of the *rank* r_k should contain especially numerous feature vectors $\mathbf{x}_j[n]$ (1) or, in other words, should have a high value of the degree of degeneration d_k (37).

There is no guarantee that the optimal vertex $\mathbf{w}_k^*[n]$ (18) resulting from the *Procedure Vertex* (34) will define the *main vertexical plane* $P_k(\mathbf{x}_{j(1)}[n],..., \mathbf{x}_{j(rk)}[n])$ (11) with the highest value of the *degree of degeneration* d_k (37). Modifications to the *Procedure Vertex* (34) could allow to find the plane $P_k(\mathbf{x}_{j(1)}[n],..., \mathbf{x}_{j(rk)}[n])$ (11) with a higher degree of degeneration d_k (37). One of these modification would be to replace the rule (35) in the step iii. with the below rule:

iii'. Find such feature vector $\mathbf{x}_{j'}[n]$ $(\mathbf{x}_{j'}[n] \in C_m)$ which when removed from the subset C_m causes the largest decrease $\Delta_{j'}(\mathbf{w}_k^*[n])$ of the minimal value $\Phi_m(\mathbf{w}_k^*[n])$ (18) of the criterion function $\Phi_m(\mathbf{w}[n])$ (16):

$$(\forall \mathbf{x}_j[n] \in C_m) \quad \Delta_{j'}(\mathbf{w}_k^*[n]) \geq \Delta_j(\mathbf{w}_k^*[n]) \qquad (38)$$

The *Procedure Vertex* (34) may allow for discovering more than one vertexical plane $P_m(\mathbf{x}_{j(1)}[n],\ldots,\mathbf{x}_{j(rk)}[n])$ (11) from a data set C (1) in subsequent cycles l. The optimal subset $C_{m(1)}^*$ is found during the first cycle ($l = 1$) of the *Procedure Vertex* (34), which began on the full data set $C_1 = C$ (1). The second cycle ($l = 2$) (34) begins on the reduced data set $C_2 = C_1 / C_{m(1)}^*$ and allows to find the optimal subset $C_{m(2)}^*$. A possible third cycle ($l = 3$) begins on the data set set $C_3 = C_2 / C_{m(2)}^*$ and allows to find the optimal subset $C_{m(3)}^*$ and so on. Subsequent cycles l allow to generate the sequence of the K optimal subset $C_{m(l)}^*$ and the optimal vertices $\mathbf{w}_{k(l)}^*[n]$ (18):

$$(C_{m(1)}^*, \mathbf{w}_{k(1)}^*[n]), (C_{m(2)}^*, \mathbf{w}_{k(2)}^*[n]),\ldots,(C_{m(K)}^*, \mathbf{w}_{k(K)}^*[n]), \tag{39}$$

As a result, the sequence of the K vertexical planes $P_{m(l)}(\mathbf{x}_{j(1)}[n],\ldots,\mathbf{x}_{j(rk)}[n])$ (11) can be generated on the basis of the optimal vertices $\mathbf{w}_{k(l)}^*[n]$ (39). The sequence of the K vertexical planes $P_{m(l)}(\mathbf{x}_{j(1)}[n],\ldots,\mathbf{x}_{j(rk)}[n])$ (11) allows among others to divide the data sets C (1) into K subsets $C(l)$ centered around K planes $P_{m(l)}(\mathbf{x}_{j(1)}[n],\ldots,\mathbf{x}_{j(rk)}[n])$ (11). Such procedure can be called as the $K - plane$ clustering.

The sequence (39) is generated through gradual reduction of the data set C (1) by successive removing of the optimal subsets $C_{m(l)}^*$. The reduced subsets $C_{m(l)}^*$ may be enlarged in successive cycles l, which could improve generalization power of the proposed procedure of the $K - plane$ clustering. For this purpose, the step ii. in the *Procedure Vertex* (34) can be replaced, by the below one with a small, positive parameter ε ($\varepsilon > 0$):

ii′. If $\Phi_{m(l)}(\mathbf{w}_k^*[n]) \le \varepsilon$, then the *Procedure Vertex* is **stopped** in the optimal
      vertex $\mathbf{w}_k^*[n]$, in the other case the next step is executed

The *Procedure Vertex* (34) with the parameter ε equal to zero ($\varepsilon = 0$) generates the optimal subset $C_{m(l)}^*$ constituted by feature vectors $\mathbf{x}_j[n]$ located precisely on the optimal plane $P_{m(l)}(\mathbf{x}_{j(1)}[n],\ldots,\mathbf{x}_{j(rk)}[n])$ (11). If the parameter ε becomes greater than zero ($\varepsilon > 0$), then the optimal subset $C_{m(l)}'$ may contain both the feature vectors $\mathbf{x}_j[n]$ located on the optimal plane $P_{m(l)}(\mathbf{x}_{j(1)}[n],\ldots,\mathbf{x}_{j(rk)}[n])$ (11) as well as near this plane:

$$C_{m(l)}' = \{\mathbf{x}_j[n]: \mathbf{x}_j[n]\in C_{m(l)} \ and \ \Phi_{m(l)}(\mathbf{w}_k^*[n]) \le \varepsilon\} \tag{40}$$

where $\Phi_{m(l)}(\mathbf{w}_k^*[n])$ is the minimal value $\Phi_{m(l)}(\mathbf{w}_k^*[n])$ (18) of the criterion function $\Phi_{m(l)}(\mathbf{w}[n])$ (16) defined on elements $\mathbf{x}_j[n]$ of the subset $C_{m(l)}$ ($C_{m(l)} \subset C$ (1)).

6   Modified *CPL* Criterion Functions with Feature Costs

The modified criterion function $\Psi_{m\lambda}(\mathbf{w}[n])$ includes additional *CPL* penalty functions $\phi_i(\mathbf{w}[n])$ in the form of the absolute values $|w_i|$:

$$(\forall i \in \{1,\ldots, n\})$$
$$\phi_i(\mathbf{w}[n]) = |w_i| = \begin{cases} -\mathbf{e}_i[n]^T\mathbf{w}[n] & if \ \ \mathbf{e}_i[n]^T\mathbf{w}[n] < 0 \\ \mathbf{e}_i[n]^T\mathbf{w}[n] & if \ \ \mathbf{e}_i[n]^T\mathbf{w}[n] \ge 0 \end{cases} \tag{41}$$

where $\mathbf{e}_i[n] = [0,\ldots,1,\ldots,0]^T$ are the unit vectors $\mathbf{e}_i[n]$.

The modified criterion function $\Psi_{m\lambda}(\mathbf{w}[n])$ is a weighted sum of the criterion function $\Phi_m(\mathbf{w}[n])$ (16) and the cost functions $\phi_i(\mathbf{w}[n])$ (41), where $i \in I = \{1,..., n\}$:

$$\Psi_{m\lambda}(\mathbf{w}[n]) = \Phi_m(\mathbf{w}[n]) + \lambda \sum_{i \in I} \gamma_i\, \phi_i(\mathbf{w}[n]) = \Phi_m(\mathbf{w}[n]) + \lambda \sum_{i \in I} \gamma_i\, |w_i| \qquad (42)$$

where λ is the *cost level* ($\lambda \geq 0$), γ_i – is the *cost* of the feature x_i ($\gamma_i > 0$), typically $\gamma_i = 1$.

Similarly as the function $\Phi_m(\mathbf{w}[n])$ (16), the modified criterion function $\Psi_{m\lambda}(\mathbf{w}[n])$ (42) is convex and piecewise linear *(CPL)*. The basis exchange algorithms allow to find efficiently the minimum $\Psi_{m\lambda}(\mathbf{w}_{k\lambda}{}^*[n])$ of the criterion function $\Psi_{m\lambda}(\mathbf{w}[n])$ (42) in one of the vertices $\mathbf{w}_k[n]$ (9) []:

$$(\exists \mathbf{w}_{k\lambda}{}^*[n])\ (\forall \mathbf{w}[n])\ \ \Psi_{m\lambda}(\mathbf{w}[n]) \geq \Psi_{m\lambda}(\mathbf{w}_{k\lambda}{}^*[n]) \qquad (43)$$

The modified criterion function $\Psi_{m\lambda}(\mathbf{w}[n])$ (42) is used in the *relaxed linear separability (RLS)* method of feature subset selection []. The reduction of unimportant features x_i in the cost sensitive manner is based in the *RLS* method on componets $w_{k,i}{}^*$ of the optimal wertex $\mathbf{w}_{k\lambda}{}^*[n] = [w_{k,1}{}^*,...,w_{k,n}{}^*]^T$ (43) [9].

$$(w_{k,i}* = 0) \Rightarrow (\text{the } i\text{-th feature } x_i \text{ is reduced}) \qquad (44)$$

The reduction of the i-th feature x_i means that the feature vectors $\mathbf{x}_j[n]$ lose their k-th component $x_{j,i}$. Such components $x_{j,i}$ can be removed without changing the location of the optimal vertex $\mathbf{w}_k{}^*[n]$ (43) or the values of inner products $\mathbf{w}_k{}^*[n]^T \mathbf{x}_j[n]$ in the criterion functions $\Phi_m(\mathbf{w}[n])$ (16) or $\Psi_{m\lambda}(\mathbf{w}[n])$ (42).

The regularization component $\lambda \sum \gamma_I\, |w_i|$ used in the function $\Psi_{m\lambda}(\mathbf{w}[n])$ (42) is similar to the one used in the *Lasso* method [10]. The *Lasso* method was developed in the framework of the regression analysis for the model selection purposes. The main difference between the *Lasso* and the *RLS* methods of feature selection is in the types of the basic criterion functions. The basic criterion function used in the *Lasso* method is usually the *Last squares* type. The basic criterion function used in the *RLS* method is the *CPL* type. This difference affects, inter alia, the computational techniques used for minimizing the criterion functions.

7   Oriented Graph G_m Based on Polytopes in Parameter Space

Feature vectors $\mathbf{x}_j[n]$ from the subset C_m ($C_m \subset C$ (1)) define the hyperplanes h_j (2) in the parameter space R^n. Similarly, n unit vectors $\mathbf{e}_i[n]$ define the hyperplanes $h_i{}^0$ (3). The hyperplanes h_j (2) and $h_i{}^0$ (3) divide the parameter space R^n into the disjoined sets *(convex polytopes)* P_l with walls, vertices $\mathbf{w}_k[n]$ (9), and edges $l_{k,k'}$ which are characterized by the below properties:

- none of the hyperplanes h_j (2) or $h_i{}^0$ (3) intersect the set P_l \qquad (45)
- each *wall* of the polytope P_l is formed by one hyperplane h_j (2) or $h_i{}^0$ (3)
- each *wertex* $\mathbf{w}_k[n]$ (9) of the set P_l is the intersection point of at least n hyperplanes h_j (2) or $h_i{}^0$ (3)
- each *edge* $l_{k,k'}$ connects two neighboring vertices $\mathbf{w}_k[n]$ and $\mathbf{w}_{k'}[n]$ of the set P_l and can be defined as the intersection of $n-1$ hyperplanes h_j (2) or $h_i{}^0$ (3)

We can remark that both the criterion function $\Phi_m(\mathbf{w}[n])$ (16) as well as the modified criterion function $\Psi_{m\lambda}(\mathbf{w}[n])$ (42) are linear inside of each polytope P_l []. The gradient $\nabla\Phi_m(\mathbf{w}[n])$ of the criterion function $\Phi_m(\mathbf{w}[n])$ (16) inside selected polytope P_l is constant and can be given by the below expression ($\mathbf{w}[n] \in P_l$):

$$\nabla\Phi_m(\mathbf{w}[n]) = \Sigma\; \alpha_j\; s_j(\mathbf{w}[n])\; \mathbf{x}_j[n] \qquad (46)$$
$$j \in J_m$$

where

$$s_j(\mathbf{w}[n]) = 1 \;\; if \;\; \mathbf{w}[n]^T\mathbf{x}_j[n] > 1 \;\; and \qquad (47)$$
$$s_j(\mathbf{w}[n]) = -1 \;\; if \;\; \mathbf{w}[n]^T\mathbf{x}_j[n] < 1$$

The gradient $\nabla\Psi_{m\lambda}(\mathbf{w}[n])$ of the modified criterion function $\Psi_{m\lambda}(\mathbf{w}[n])$ (42) can be specified in a similar manner in a point $\mathbf{w}[n] = [w_1,\ldots,w_n]^T$ from the polytope P_l:

$$\nabla\Psi_{m\lambda}(\mathbf{w}[n]) = \nabla\Phi_m(\mathbf{w}[n]) + \lambda \Sigma\; \gamma_i\; s_i^0(\mathbf{w}[n])\; \mathbf{e}_i[n] \qquad (48)$$
$$i \in I$$

where

$$s_j^0(\mathbf{w}[n]) = 1 \;\; if \;\; w_i > 0 \;\; and \qquad (49)$$
$$s_j^0(\mathbf{w}[n]) = -1 \;\; if \;\; w_i < 0$$

Remark 5: The gradient $\nabla\Psi_{m\lambda}(\mathbf{w}[n])$ of the modified criterion function $\Psi_{m\lambda}(\mathbf{w}[n])$ (42) can be reduced to the gradient $\nabla\Phi_m(\mathbf{w}[n])$ (46) of the criterion function $\Phi_m(\mathbf{w}[n])$ inside each polytope P_l by reducing the cost level λ to zero.

The gradient $\nabla\Psi_{m\lambda}(\mathbf{w}[n])$ (48) of the modified criterion function $\Psi_{m\lambda}(\mathbf{w}[n])$ (42) allows for the below orientation of the edges $l_{k,k'}$ connecting two neighboring vertices $\mathbf{w}_k[n]$ and $\mathbf{w}_{k'}[n]$ of particular polytopes P_l:

$$l_{k,k'} = \mathbf{w}_{k'}[n] - \mathbf{w}_k[n] \;\; if \;\; \nabla\Psi_{m\lambda}(\mathbf{w}[n])^T(\mathbf{w}_{k'}[n] - \mathbf{w}_k[n]) < 0 \;\; and \qquad (50)$$
$$l_{k,k'} = \mathbf{w}_k[n] - \mathbf{w}_{k'}[n] \;\; if \;\; \nabla\Psi_{m\lambda}(\mathbf{w}[n])^T(\mathbf{w}_k[n] - \mathbf{w}_{k'}[n]) < 0$$

Remark 4: Each edge $l_{k,k'}$ is oriented (50) decreasingly to the criterion function $\Psi_{m\lambda}(\mathbf{w}[n])$ (42). This means that the move between two vertices $\mathbf{w}_k[n]$ and $\mathbf{w}_{k'}[n]$ in accordance with the edge $l_{k,k'}$ (50) always causes a decrease of the function $\Psi_{m\lambda}(\mathbf{w}[n])$.

The oriented graph G_m is defined on the basis of the set of the vertices $\{\mathbf{w}_k[n]\}$ (9) and the set of the oriented edges $\{l_{k,k'}\}$ (50):

$$G_m = (\{\mathbf{w}_k[n]\}, \{l_{k,k'}\}) \qquad (51)$$

Remark 5: The oriented graph G_m (51) has no loops [8].

Any algorithm based on moving between vertices $\mathbf{w}_k[n]$ of the graph G (52) in accordance with the edges $l_{k,k'}$ orientation (50) reaches the optimal vertex $\mathbf{w}_{k\lambda}^*[n]$ (43) after a finite number of steps. The oriented graph G_m (51) with the above properties has been used in the proof of the basis exchange algorithm convergence in a finite number of steps []. These types of algorithms was been used among others for the minimization of the criterion functions $\Phi_m(\mathbf{w}[n])$ (16) or $\Psi_{m\lambda}(\mathbf{w}[n])$ (42).

The modified criterion function $\Psi_{m\lambda}(\mathbf{w}[n])$ (42) can be used for the purpose of reducing the *rank* r_k (*Definition* 1) of the optimal vertices $\mathbf{w}_{k\lambda}^*[n]$ (43). We can infer on the basis of the formula (42) that an increase of the the the *cost level* λ causes an increase of the number of the components $w_{k,i}^*$ of the vector $\mathbf{w}_{k\lambda}^*[n]$ (43) equal to zero ($w_{k,i}^* = 0$). Therefore we can arbitrarily reduce the rank r_k of the optimal vertex $\mathbf{w}_{k\lambda}^*[n]$ (43) by choosing a sufficiently large value of the parameter λ (42).

Supposition II: Discovering *main vertexical planes* $P_k(\mathbf{x}_{j(1)}[n],\ldots, \mathbf{x}_{j(rk)}[n])$ (11) (12) in the feature space $F[n]$ can be based on the detection and examination of vertices $\mathbf{w}_k[n]$ of the oriented graph G_m (51) with a large number of edges $l_{k,k'}$.

Discovering *"flat patterns"* located on the *main vertexical planes* $P_k(\mathbf{x}_{i(1)}[n],\ldots, \mathbf{x}_{i(rk)}[n])$ (11) (12) of different ranks r_k allows to design different models of linear interactions between features x_i or objects $\mathbf{x}_i[n]$. Different linear models of interactions (relations) between features x_i or objects $\mathbf{x}_i[n]$ of a given *"flat pattern"* can be determined by using the base matrices $\mathbf{B}_k[n]$ (9) linked to vertices $\mathbf{w}_k[n]$ (9) of different ranks r_k.

8   An Example - A Toy Data Set in a Two-Dimensional Feature Space

Feature vectors $\mathbf{x}_j[n]$ situated on the main vertexical plane $P_k(\mathbf{x}_{j(1)}[n],\ldots, \mathbf{x}_{j(rk)}[n])$ (11) (12) define the hyperplanes h_j (2) passing through the vertex $\mathbf{w}_k[n]$ (9) of the rank r_k, which is characterized by a high degree of degeneration d_k. Such vertex $\mathbf{w}_k[n]$ in the graph G_m (51) is characterized by a large number of the oriented edges $\{l_{k,k'}\}$ (50).

To illustrate this property the artificial data sets shown in the *Table* 1 have been used. The *Table* 1 contains feature vectors $\mathbf{x}_j[2]$ situated along three lines (*Figure* 1). The resulting graph G_m (51) contains three degenerated vertices (*Figure* 2).

Table 1. The artificial data sets *Line* I, *Line* II and *Line* III of two-dimensional feature vectors $\mathbf{x}_j[2] = [x_{j,1}, x_{j,2}]^T$ ($\mathbf{x}_j[2] \in R^2$)

Number j	Line I $[x_{j,1}, x_{j,2}]$	Line II $[x_{j,1}, x_{j,2}]$	Line III $[x_{j,1}, x_{j,2}]$
1	[-1, 1]	[2, 1]	[3, -2]
2	[0, 1]	[0, -1]	[1, 0]
3	[1, 1]	[1, 0]	[0, 1]

The sets from the *Table* 1 are represented on the *Figure* 1 and the *Figure* 2.

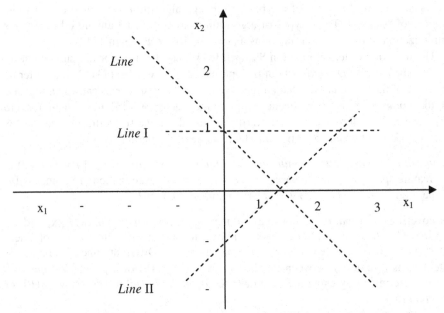

Fig. 1. Vertexical planes (lines) $P_m(\mathbf{x}_{j(1)}[2],\mathbf{x}_{j(2)}[2])$ (11) generated by the data sets *Line* I, *Line* II and *Line* III from the *Table* 1

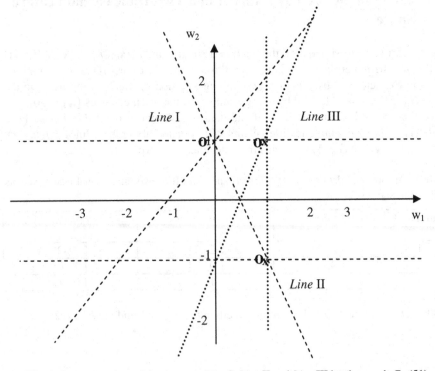

Fig. 2. Representation of the data sets *Line* I, *Line* II and *Line* III by the graph G_m (51)

9   Concluding Remarks

The main vertexical planes $P_k(x_{j(1)}[n],\ldots, x_{j(rk)}[n])$ (11) (12) contain *"flat patterns"* in the feature space $F[n]$. The proposed procedure (34) of vertexical planes discovering from multivariate data set C (1) is based on the multiple minimization of the of the criterion function $\Phi_m(w[n])$ (16).

The modified criterion function $\Psi_{m\lambda}(w[n])$ (42) allows to reduce dimensionality $r_k - 1$ of the main vertexical planes $P_k(x_{j(1)}[n],\ldots, x_{j(rk)}[n])$ (11) (12). Such planes with varied dimensionality could be useful, among others, in creating different degree models of interaction between features x_i or objects $x_j[n]$.

The method of linking the main vertexical planes $P_k(x_{j(1)}[n],\ldots, x_{j(rk)}[n])$ (11) (12) in the feature space $F[n]$ with the degenerated vertices (9) of the oriented graphs G_m (51) is also described in the paper. This relationship which is based on common vertices $w_k[n]$ (9) could create a bridge between data mining methods and graph methods. Such relationship could be used, for example, in modeling of social networks or in the decomposition of *mixture models* (e.g. [11]).

Acknowledgments. This work was supported by the project S/WI/2/2014 from the Białystok University of Technology, Poland.

References

1. Johnson, R.A., Wichern, D.W.: Applied Multivariate Statistical Analysis. Prentice-Hall, Inc., Englewood Cliffs (1991)
2. Duda, O.R., Hart, P.E., Stork, D.G.: Pattern Classification. J. Wiley, New York (2001)
3. Hand, D., Smyth, P., Mannila, H.: Principles of data mining. MIT Press, Cambridge (2001)
4. Bobrowski, L., Bezdek, J.C.: C-means clustering with the L_1 and L_∞ norms. IEEE Transactions on Systems Man and Cybernetics 21(3), 545–554 (1991a)
5. Bradley, P.S., Mangasarian, O.L.: K-plane clustering. Journal of Global Optimization 16, 23–32 (2000)
6. Bobrowski, L.: K-lines clustering with convex and piecewise linear (CPL) functions. In: MATHMOD, Vienna (2012)
7. Bobrowski, L.: Data mining based on convex and piecewise linear criterion functions. Technical University Białystok (2005) (in Polish)
8. Bobrowski, L.: Design of piecewise linear classifiers from formal neurons by some basis exchange technique. Pattern Recognition 24(9), 863–870 (1991)
9. Bobrowski, L., Łukaszuk, T.: Relaxed Linear Separability (*RLS*) Approach to Feature (Gene) Subset Selection. In: Xia, X. (ed.) Selected Works in Bioinformatics, pp. 103–118. INTECH (2011)
10. Tibshirani, R.: Regression shrinkage and selection via the lasso. Journal of the Royal Statistical Society, Series B 58(1), 267–288 (1996)
11. Kwedlo, W.: A New Method for Random Initialization of the EM Algorithm for Multivariate Gaussian Mixture Learning. In: Burduk, R., Jackowski, K., Kurzynski, M., Wozniak, M., Zolnierek, A. (eds.) CORES 2013. AISC, vol. 226, pp. 85–94. Springer, Heidelberg (2013)

Big Data Analytics: A Literature Review Paper

Nada Elgendy and Ahmed Elragal

Department of Business Informatics & Operations,
German University in Cairo (GUC), Cairo, Egypt
{nada.el-gendy,ahmed.elragal}@guc.edu.eg

Abstract. In the information era, enormous amounts of data have become available on hand to decision makers. Big data refers to datasets that are not only big, but also high in variety and velocity, which makes them difficult to handle using traditional tools and techniques. Due to the rapid growth of such data, solutions need to be studied and provided in order to handle and extract value and knowledge from these datasets. Furthermore, decision makers need to be able to gain valuable insights from such varied and rapidly changing data, ranging from daily transactions to customer interactions and social network data. Such value can be provided using big data analytics, which is the application of advanced analytics techniques on big data. This paper aims to analyze some of the different analytics methods and tools which can be applied to big data, as well as the opportunities provided by the application of big data analytics in various decision domains.

Keywords: big data, data mining, analytics, decision making.

1 Introduction

Imagine a world without data storage; a place where every detail about a person or organization, every transaction performed, or every aspect which can be documented is lost directly after use. Organizations would thus lose the ability to extract valuable information and knowledge, perform detailed analyses, as well as provide new opportunities and advantages. Anything ranging from customer names and addresses, to products available, to purchases made, to employees hired, etc. has become essential for day-to-day continuity. Data is the building block upon which any organization thrives.

Now think of the extent of details and the surge of data and information provided nowadays through the advancements in technologies and the internet. With the increase in storage capabilities and methods of data collection, huge amounts of data have become easily available. Every second, more and more data is being created and needs to be stored and analyzed in order to extract value. Furthermore, data has become cheaper to store, so organizations need to get as much value as possible from the huge amounts of stored data.

The size, variety, and rapid change of such data require a new type of big data analytics, as well as different storage and analysis methods. Such sheer amounts of big data need to be properly analyzed, and pertaining information should be extracted.

P. Perner (Ed.): ICDM 2014, LNAI 8557, pp. 214–227, 2014.
© Springer International Publishing Switzerland 2014

The contribution of this paper is to provide an analysis of the available literature on big data analytics. Accordingly, some of the various big data tools, methods, and technologies which can be applied are discussed, and their applications and opportunities provided in several decision domains are portrayed.

The literature was selected based on its novelty and discussion of important topics related to big data, in order to serve the purpose of our research. The publication years range from 2008-2013, with most of the literature focusing on big data ranging from 2011-2013. This is due to big data being a recently focused upon topic. Furthermore, our corpus mostly includes research from some of the top journals, conferences, and white papers by leading corporations in the industry. Due to long review process of journals, most of the papers discussing big data analytics, its tools and methods, and its applications were found to be conference papers, and white papers. While big data analytics is being researched in academia, several of the industrial advancements and new technologies provided were mostly discussed in industry papers.

2   Big Data Analytics

The term "Big Data" has recently been applied to datasets that grow so large that they become awkward to work with using traditional database management systems. They are data sets whose size is beyond the ability of commonly used software tools and storage systems to capture, store, manage, as well as process the data within a tolerable elapsed time [12].

Big data sizes are constantly increasing, currently ranging from a few dozen terabytes (TB) to many petabytes (PB) of data in a single data set. Consequently, some of the difficulties related to big data include capture, storage, search, sharing, analytics, and visualizing. Today, enterprises are exploring large volumes of highly detailed data so as to discover facts they didn't know before [17].

Hence, big data analytics is where advanced analytic techniques are applied on big data sets. Analytics based on large data samples reveals and leverages business change. However, the larger the set of data, the more difficult it becomes to manage [17].

In this section, we will start by discussing the characteristics of big data, as well as its importance. Naturally, business benefit can commonly be derived from analyzing larger and more complex data sets that require real time or near-real time capabilities; however, this leads to a need for new data architectures, analytical methods, and tools. Therefore the successive section will elaborate the big data analytics tools and methods, in particular, starting with the big data storage and management, then moving on to the big data analytic processing. It then concludes with some of the various big data analyses which have grown in usage with big data.

2.1   Characteristics of Big Data

Big data is data whose scale, distribution, diversity, and/or timeliness require the use of new technical architectures, analytics, and tools in order to enable insights that unlock new sources of business value. Three main features characterize big data: volume, variety, and velocity, or the three V's. The volume of the data is its size, and how enormous it is. Velocity refers to the rate with which data is changing, or how often it is created. Finally, variety includes the different formats and types of data, as well as the different kinds of uses and ways of analyzing the data [9].

Data volume is the primary attribute of big data. Big data can be quantified by size in TBs or PBs, as well as even the number of records, transactions, tables, or files. Additionally, one of the things that make big data really big is that it's coming from a greater variety of sources than ever before, including logs, clickstreams, and social media. Using these sources for analytics means that common structured data is now joined by unstructured data, such as text and human language, and semi-structured data, such as eXtensible Markup Language (XML) or Rich Site Summary (RSS) feeds. There's also data, which is hard to categorize since it comes from audio, video, and other devices. Furthermore, multi-dimensional data can be drawn from a data warehouse to add historic context to big data. Thus, with big data, variety is just as big as volume.

Moreover, big data can be described by its velocity or speed. This is basically the frequency of data generation or the frequency of data delivery. The leading edge of big data is streaming data, which is collected in real-time from the websites [17]. Some researchers and organizations have discussed the addition of a fourth V, or veracity. Veracity focuses on the quality of the data. This characterizes big data quality as good, bad, or undefined due to data inconsistency, incompleteness, ambiguity, latency, deception, and approximations [22].

2.2   Big Data Analytics Tools and Methods

With the evolution of technology and the increased multitudes of data flowing in and out of organizations daily, there has become a need for faster and more efficient ways of analyzing such data. Having piles of data on hand is no longer enough to make efficient decisions at the right time.

Such data sets can no longer be easily analyzed with traditional data management and analysis techniques and infrastructures. Therefore, there arises a need for new tools and methods specialized for big data analytics, as well as the required architectures for storing and managing such data. Accordingly, the emergence of big data has an effect on everything from the data itself and its collection, to the processing, to the final extracted decisions.

Consequently, [8] proposed the Big – Data, Analytics, and Decisions (B-DAD) framework which incorporates the big data analytics tools and methods into the decision making process [8]. The framework maps the different big data storage, management, and processing tools, analytics tools and methods, and visualization and evaluation tools to the different phases of the decision making process. Hence, the changes associated with big data analytics are reflected in three main areas: big data storage and architecture, data and analytics processing, and, finally, the big data analyses which can be applied for knowledge discovery and informed decision making. Each area will be further discussed in this section. However, since big data is still evolving as an important field of research, and new findings and tools are constantly developing, this section is not exhaustive of all the possibilities, and focuses on providing a general idea, rather than a list of all potential opportunities and technologies.

Big Data Storage and Management

One of the first things organizations have to manage when dealing with big data, is where and how this data will be stored once it is acquired. The traditional methods of structured data storage and retrieval include relational databases, data marts, and data warehouses. The data is uploaded to the storage from operational data stores using Extract, Transform, Load (ETL), or Extract, Load, Transform (ELT), tools which extract the data from outside sources, transform the data to fit operational needs, and finally load the data into the database or data warehouse. Thus, the data is cleaned, transformed, and catalogued before being made available for data mining and online analytical functions [3].

However, the big data environment calls for Magnetic, Agile, Deep (MAD) analysis skills, which differ from the aspects of a traditional Enterprise Data Warehouse (EDW) environment. First of all, traditional EDW approaches discourage the incorporation of new data sources until they are cleansed and integrated. Due to the ubiquity of data nowadays, big data environments need to be magnetic, thus attracting all the data sources, regardless of the data quality [5]. Furthermore, given the growing numbers of data sources, as well as the sophistication of the data analyses, big data storage should allow analysts to easily produce and adapt data rapidly. This requires an agile database, whose logical and physical contents can adapt in sync with rapid data evolution [11]. Finally, since current data analyses use complex statistical methods, and analysts need to be able to study enormous datasets by drilling up and down, a big data repository also needs to be deep, and serve as a sophisticated algorithmic runtime engine [5].

Accordingly, several solutions, ranging from distributed systems and Massive Parallel Processing (MPP) databases for providing high query performance and platform scalability, to non-relational or in-memory databases, have been used for big data.

Non-relational databases, such as Not Only SQL (NoSQL), were developed for storing and managing unstructured, or non-relational, data. NoSQL databases aim for massive scaling, data model flexibility, and simplified application development and deployment. Contrary to relational databases, NoSQL databases separate data management and data storage. Such databases rather focus on the high-performance scalable data storage, and allow data management tasks to be written in the application layer instead of having it written in databases specific languages [3].

On the other hand, in-memory databases manage the data in server memory, thus eliminating disk input/output (I/O) and enabling real-time responses from the database. Instead of using mechanical disk drives, it is possible to store the primary database in silicon-based main memory. This results in orders of magnitude of improvement in the performance, and allows entirely new applications to be developed [16]. Furthermore, in-memory databases are now being used for advanced analytics on big data, especially to speed the access to and scoring of analytic models for analysis. This provides scalability for big data, and speed for discovery analytics [17].

Alternatively, Hadoop is a framework for performing big data analytics which provides reliability, scalability, and manageability by providing an implementation for the MapReduce paradigm, which is discussed in the following section, as well as gluing the storage and analytics together. Hadoop consists of two main components: the HDFS for the big data storage, and MapReduce for big data analytics [9]. The HDFS storage function provides a redundant and reliable distributed file system, which is optimized for large files, where a single file is split into blocks and distributed across cluster nodes. Additionally, the data is protected among the nodes by a replication mechanism, which ensures availability and reliability despite any node failures [3]. There are two types of HDFS nodes: the Data Nodes and the Name Nodes. Data is stored in replicated file blocks across the multiple Data Nodes, and the Name Node acts as a regulator between the client and the Data Node, directing the client to the particular Data Node which contains the requested data [3].

Big Data Analytic Processing

After the big data storage, comes the analytic processing. According to [10], there are four critical requirements for big data processing. The first requirement is fast data loading. Since the disk and network traffic interferes with the query executions during data loading, it is necessary to reduce the data loading time. The second requirement is fast query processing. In order to satisfy the requirements of heavy workloads and real-time requests, many queries are response-time critical. Thus, the data placement structure must be capable of retaining high query processing speeds as the amounts of queries rapidly increase. Additionally, the third requirement for big data processing is the highly efficient utilization of storage space. Since the rapid growth in user activities can demand scalable storage capacity and computing power, limited disk space necessitates that data storage be well managed during processing, and issues on how to store the data so that space utilization is maximized be addressed. Finally, the fourth requirement is the strong adaptivity to highly dynamic workload patterns. As big data sets are analyzed by different applications and users, for different purposes, and in various ways, the underlying system should be highly adaptive to unexpected dynamics in data processing, and not specific to certain workload patterns [10].

Map Reduce is a parallel programming model, inspired by the "Map" and "Reduce" of functional languages, which is suitable for big data processing. It is the core of Hadoop, and performs the data processing and analytics functions [6]. According to EMC, the MapReduce paradigm is based on adding more computers or resources, rather than increasing the power or storage capacity of a single computer; in other words, scaling out rather than scaling up [9]. The fundamental idea of MapReduce is breaking a task down into stages and executing the stages in parallel in order to reduce the time needed to complete the task [6].

The first phase of the MapReduce job is to map input values to a set of key/value pairs as output. The "Map" function accordingly partitions large computational tasks into smaller tasks, and assigns them to the appropriate key/value pairs [6]. Thus, unstructured data, such as text, can be mapped to a structured key/value pair, where, for example, the key could be the word in the text and the value is the number of occurrences of the word. This output is then the input to the "Reduce" function [9]. Reduce then performs the collection and combination of this output, by combining all values which share the same key value, to provide the final result of the computational task [6].

The MapReduce function within Hadoop depends on two different nodes: the Job Tracker and the Task Tracker nodes. The Job Tracker nodes are the ones which are responsible for distributing the mapper and reducer functions to the available Task Trackers, as well as monitoring the results [9]. The MapReduce job starts by the Job-Tracker assigning a portion of an input file on the HDFS to a map task, running on a node [13]. On the other hand, the Task Tracker nodes actually run the jobs and communicate results back to the Job Tracker. That communication between nodes is often through files and directories in HDFS, so inter-node communication is minimized [9].

Figure 1 shows how the MapReduce nodes and the HDFS work together. At step 1, there is a very large dataset including log files, sensor data, or anything of the sorts. The HDFS stores replicas of the data, represented by the blue, yellow, beige, and pink icons, across the Data Nodes. In step 2, the client defines and executes a map job and a reduce job on a particular data set, and sends them both to the Job Tracker. The Job Tracker then distributes the jobs across the Task Trackers in step 3. The Task Tracker runs the mapper, and the mapper produces output that is then stored in the HDFS file system. Finally, in step 4, the reduce job runs across the mapped data in order to produce the result.

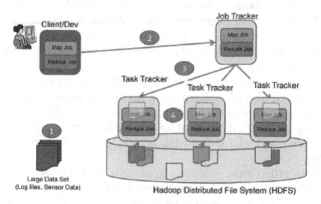

Fig. 1. MapReduce and HDFS

Hadoop is a MAD system, thus making it popular for big data analytics by loading data as files into the distributed file system, and running parallel MapReduce computations on the data. Hadoop gets its magnetism and agility from the fact that data is loaded into Hadoop simply by copying files into the distributed file system, and MapReduce interprets the data at processing time rather than loading time [11]. Thus, it is capable of attracting all data sources, as well as adapting its engines to any evolutions that may occur in such big data sources [6].

After big data is stored, managed, and processed, decision makers need to extract useful insights by performing big data analyses. In the subsections below, various big data analyses will be discussed, starting with selected traditional advanced data analytics methods, and followed by examples of some of the additional, applicable big data analyses.

Big Data Analytics

Nowadays, people don't just want to collect data, they want to understand the meaning and importance of the data, and use it to aid them in making decisions. Data analytics is the process of applying algorithms in order to analyze sets of data and extract useful and unknown patterns, relationships, and information [1]. Furthermore, data analytics are used to extract previously unknown, useful, valid, and hidden patterns and information from large data sets, as well as to detect important relationships among the stored variables. Therefore, analytics have had a significant impact on research and technologies, since decision makers have become more and more interested in learning from previous data, thus gaining competitive advantage [21].

Along with some of the most common advanced data analytics methods, such as association rules, clustering, classification and decision trees, and regression some additional analyses have become common with big data.

For example, social media has recently become important for social networking and content sharing. Yet, the content that is generated from social media websites is enormous and remains largely unexploited. However, social media analytics can be used to analyze such data and extract useful information and predictions [2]. Social media analytics is based on developing and evaluating informatics frameworks and tools in order to collect, monitor, summarize, analyze, as well as visualize social media data. Furthermore, social media analytics facilitates understanding the reactions and conversations between people in online communities, as well as extracting useful patterns and intelligence from their interactions, in addition to what they share on social media websites [24].

On the other hand, Social Network Analysis (SNA) focuses on the relationships among social entities, as well as the patterns and implications of such relationships [23]. An SNA maps and measures both formal and informal relationships in order to comprehend what facilitates the flow of knowledge between interacting parties, such as who knows who, and who shares what knowledge or information with who and using what [19].

However, SNA differs from social media analysis, in that SNA tries to capture the social relationships and patterns between networks of people. On the other hand, social media analysis aims to analyze what social media users are saying in order to uncover useful patterns, information about the users, and sentiments. This is traditionally done using text mining or sentiment analysis, which are discussed below.

On the other hand, text mining is used to analyze a document or set of documents in order to understand the content within and the meaning of the information contained. Text mining has become very important nowadays since most of the information stored, not including audio, video, and images, consists of text. While data mining deals with structured data, text presents special characteristics which basically follow a non-relational form [18].

Moreover, sentiment analysis, or opinion mining, is becoming more and more important as online opinion data, such as blogs, product reviews, forums, and social data from social media sites like Twitter and Facebook, grow tremendously. Sentiment analysis focuses on analyzing and understanding emotions from subjective text patterns, and is enabled through text mining. It identifies opinions and attitudes of individuals towards certain topics, and is useful in classifying viewpoints as positive or negative. Sentiment analysis uses natural language processing and text analytics in order to identify and extract information by finding words that are indicative of a sentiment, as well as relationships between words, so that sentiments can be accurately identified [15].

Finally, from the strongest potential growths among big data analytics options is Advanced Data Visualization (ADV) and visual discovery [17]. Presenting information so that people can consume it effectively is a key challenge that needs to be met, in order for decision makers to be able to properly analyze data in a way to lead to concrete actions [14].

ADV has emerged as a powerful technique to discover knowledge from data. ADV combines data analysis methods with interactive visualization to enable comprehensive data exploration. It is a data driven exploratory approach that fits well in situations where analysts have little knowledge about the data [20]. With the generation of more and more data of high volume and complexity, an increasing demand has arisen for ADV solutions from many application domains [25]. Additionally, such visualization analyses take advantage of human perceptual and reasoning abilities, which enables them to thoroughly analyze data at both the overview and the detailed levels. Along with the size and complexity of big data, intuitive visual representation and interaction is needed to facilitate the analyst's perception and reasoning [20].

ADV can enable faster analysis, better decision making, and more effective presentation and comprehension of results by providing interactive statistical graphics and a point-and-click interface [4]. Furthermore, ADV is a natural fit for big data since it can scale its visualizations to represent thousands or millions of data points, unlike standard pie, bar, and line charts. Moreover, it can handle diverse data types, as well as present analytic data structures that aren't easily flattened onto a computer screen, such as hierarchies and neural nets. Additionally, most ADV tools and functions can support interfaces to all the leading data sources, thus enabling business analysts to explore data widely across a variety of sources in search of the right analytics dataset, usually in real-time [17].

3   Big Data Analytics and Decision Making

From the decision maker's perspective, the significance of big data lies in its ability to provide information and knowledge of value, upon which to base decisions. The managerial decision making process has been an important and thoroughly covered topic in research throughout the years.

Big data is becoming an increasingly important asset for decision makers. Large volumes of highly detailed data from various sources such as scanners, mobile phones, loyalty cards, the web, and social media platforms provide the opportunity to deliver significant benefits to organizations. This is possible only if the data is properly analyzed to reveal valuable insights, allowing for decision makers to capitalize upon the resulting opportunities from the wealth of historic and real-time data generated through supply chains, production processes, customer behaviors, etc. [4].

Moreover, organizations are currently accustomed to analyzing internal data, such as sales, shipments, and inventory. However, the need for analyzing external data, such as customer markets and supply chains, has arisen, and the use of big data can provide cumulative value and knowledge. With the increasing sizes and types of unstructured data on hand, it becomes necessary to make more informed decisions based on drawing meaningful inferences from the data [7].

Accordingly, [8] developed the B-DAD framework which maps big data tools and techniques, into the decision making process [8]. Such a framework is intended to enhance the quality of the decision making process in regards to dealing with big data. The first phase of the decision making process is the intelligence phase, where data which can be used to identify problems and opportunities is collected from internal and external data sources. In this phase, the sources of big data need to be identified, and the data needs to be gathered from different sources, processed, stored, and migrated to the end user. Such big data needs to be treated accordingly, so after the data sources and types of data required for the analysis are defined, the chosen data is acquired and stored in any of the big data storage and management tools previously discussed After the big data is acquired and stored, it is then organized, prepared, and processed, This is achieved across a high-speed network using ETL/ELT or big data processing tools, which have been covered in the previous sections.

The next phase in the decision making process is the design phase, where possible courses of action are developed and analyzed through a conceptualization, or a representative model of the problem. The framework divides this phase into three steps, model planning, data analytics, and analyzing. Here, a model for data analytics, such as those previously discussed, is selected and planned, and then applied, and finally analyzed.

Consequently, the following phase in the decision making process is the choice phase, where methods are used to evaluate the impacts of the proposed solutions, or courses of action, from the design phase. Finally, the last phase in the decision making process is the implementation phase, where the proposed solution from the previous phase is implemented [8].

As the amount of big data continues to exponentially grow, organizations throughout the different sectors are becoming more interested in how to manage and analyze such data. Thus, they are rushing to seize the opportunities offered by big data, and gain the most benefit and insight possible, consequently adopting big data analytics in order to unlock economic value and make better and faster decisions. Therefore, organizations are turning towards big data analytics in order to analyze huge amounts of data faster, and reveal previously unseen patterns, sentiments, and customer intelligence. This section focuses on some of the different applications, both proposed and implemented, of big data analytics, and how these applications can aid organizations across different sectors to gain valuable insights and enhance decision making.

According to Manyika et al.'s research, big data can enable companies to create new products and services, enhance existing ones, as well as invent entirely new business models. Such benefits can be gained by applying big data analytics in different areas, such as customer intelligence, supply chain intelligence, performance, quality and risk management and fraud detection [14]. Furthermore, Cebr's study highlighted the main industries that can benefit from big data analytics, such as the manufacturing, retail, central government, healthcare, telecom, and banking industries [4].

3.1   Customer Intelligence

Big data analytics holds much potential for customer intelligence, and can highly benefit industries such as retail, banking, and telecommunications. Big data can create transparency, and make relevant data more easily accessible to stakeholders in a timely manner [14]. Big data analytics can provide organizations with the ability to profile and segment customers based on different socioeconomic characteristics, as well as increase levels of customer satisfaction and retention [4]. This can allow them to make more informed marketing decisions, and market to different segments based on their preferences along with the recognition of sales and marketing opportunities [17]. Moreover, social media can be used to inform companies what their customers like, as well as what they don't like. By performing sentiment analysis on this data, firms can be alerted beforehand when customers are turning against them or shifting to different products, and accordingly take action [7].

Additionally, using SNAs to monitor customer sentiments towards brands, and identify influential individuals, can help organizations react to trends and perform direct marketing. Big data analytics can also enable the construction of predictive models for customer behavior and purchase patterns, therefore raising overall profitability [4]. Even organizations which have used segmentation for many years are beginning to deploy more sophisticated big data techniques, such as real-time micro-segmentation of customers, in order to target promotions and advertising [14]. Consequently, big data analytics can benefit organizations by enabling better targeted social influencer marketing, defining and predicting trends from market sentiments, as well as analyzing and understanding churn and other customer behaviors [17].

3.2   Supply Chain and Performance Management

As for supply chain management, big data analytics can be used to forecast demand changes, and accordingly match their supply. This can increasingly benefit the manufacturing, retail, as well as transport and logistics industries. By analyzing stock utilization and geospatial data on deliveries, organizations can automate replenishment decisions, which will reduce lead times and minimize costs and delays, as well as process interruptions. Additionally, decisions on changing suppliers, based on quality or price competitiveness, can be taken by analyzing supplier data to monitor performance. Furthermore, alternate pricing scenarios can be run instantly, which can enable a reduction in inventories and an increase in profit margins [4]. Accordingly, big data can lead to the identification of the root causes of cost, and provide for better planning and forecasting [17].

Another area where big data analytics can be of value is performance management, where the governmental and healthcare industries can easily benefit. With the increasing need to improve productivity, staff performance information can be monitored and forecasted by using predictive analytics tools. This can allow departments to link their strategic objectives with the service or user outcomes, thus leading to increased efficiencies. Additionally, with the availability of big data and performance information, as well as its accessibility to operations managers, the use of predictive KPIs, balanced scorecards, and dashboards within the organization can introduce operational benefits by enabling the monitoring of performance, as well as improving transparency, objectives setting, and planning and management functions [4].

3.3   Quality Management and Improvement

Especially for the manufacturing, energy and utilities, and telecommunications industries, big data can be used for quality management, in order to increase profitability and reduce costs by improving the quality of goods and services provided. For example, in the manufacturing process, predictive analytics on big data can be used to minimize the performance variability, as well as prevent quality issues by providing early warning alerts. This can reduce scrap rates, and decrease the time to market, since identifying any disruptions to the production process before they occur can save significant expenditures [4]. Additionally, big data analytics can result in manufacturing lead improvements [17]. Furthermore, real-time data analyses and monitoring of machine logs can enable managers to make swifter decisions for quality management. Also, big data analytics can allow for the real-time monitoring of network demand, in addition to the forecasting of bandwidth in response to customer behavior.

Moreover, healthcare IT systems can improve the efficiency and quality of care, by communicating and integrating patient data across different departments and institutions, while retaining privacy controls [4]. Analyzing electronic health records can improve the continuity of care for individuals, as well as creating a massive dataset through which treatments and outcomes can be predicted and compared. Therefore, with the increasing use of electronic health records, along with the advancements in analytics tools, there arises an opportunity to mine the available de-identified patient information for assessing the quality of healthcare, as well as managing diseases and health services [22].

Additionally, the quality of citizens' lives can be improved through the utilization of big data. For healthcare, sensors can be used in hospitals and homes to provide the continuous monitoring of patients, and perform real-time analyses on the patient data streaming in. This can be used to alert individuals and their health care providers if any health anomalies are detected in the analysis, requiring the patient to seek medical help [22]. Patients can also be monitored remotely to analyze their adherence to their prescriptions, and improve drug and treatment options [14].

Moreover, by analyzing information from distributed sensors on handheld devices, roads, and vehicles, which provide real-time traffic information, transportation can be transformed and improved. Traffic jams can be predicted and prevented, and drivers can operate more safely and with less disruption to the traffic flow. Such a new type of traffic ecosystem, with "intelligent" connected cars, can potentially renovate transportation and how roadways are used [22]. Accordingly, big data applications can provide smart routing, according to real-time traffic information based on personal location data. Furthermore, such applications can automatically call for help when trouble is detected by the sensors, and inform users about accidents, scheduled roadwork, and congested areas in real-time [14].

Furthermore, big data can be used for better understanding changes in the location, frequency, and intensity of weather and climate. This can benefit citizens and businesses that rely upon weather, such as farmers, as well as tourism and transportation companies. Also, with new sensors and analysis techniques for developing long term climate models and nearer weather forecasts, weather related natural disasters can be predicted, and preventive or adaptive measures can be taken beforehand [22].

3.4   Risk Management and Fraud Detection

Industries such as investment or retail banking, as well as insurance, can benefit from big data analytics in the area of risk management. Since the evaluation and bearing of risk is a critical aspect for the financial services sector, big data analytics can help in selecting investments by analyzing the likelihood of gains against the likelihood of losses. Additionally, internal and external big data can be analyzed for the full and dynamic appraisal of risk exposures [4]. Accordingly, big data can benefit organizations by enabling the quantification of risks [17]. High-performance analytics can also be used to integrate the risk profiles managed in isolation across separate departments, into enterprise wide risk profiles. This can aid in risk mitigation, since a comprehensive view of the different risk types and their interrelations is provided to decision makers [4].

Furthermore, new big data tools and technologies can provide for managing the exponential growth in network produced data, as well reduce database performance problems by increasing the ability to scale and capture the required data. Along with the enhancement in cyber analytics and data intensive computing solutions, organizations can incorporate multiple streams of data and automated analyses to protect themselves against cyber and network attacks [22].

As for fraud detection, especially in the government, banking, and insurance industries, big data analytics can be used to detect and prevent fraud [17]. Analytics are already commonly used in automated fraud detection, but organizations and sectors are looking towards harnessing the potentials of big data in order to improve their systems. Big data can allow them to match electronic data across several sources, between both public and private sectors, and perform faster analytics [4].

In addition, customer intelligence can be used to model normal customer behavior, and detect suspicious or divergent activities through the accurate flagging of outlier occurrences. Furthermore, providing systems with big data about prevailing fraud patterns can allow these systems to learn the new types of frauds and act accordingly, as the fraudsters adapt to the old systems designed to detect them. Also, SNAs can be used to identify the networks of collaborating fraudsters, as well as discover evidence of fraudulent insurance or benefits claims, which will lead to less fraudulent activity going undiscovered [4]. Thus, big data tools, techniques, and governance processes can increase the prevention and recovery of fraudulent transactions by dramatically increasing the speed of identification and detection of compliance patterns within all available data sets [22].

4   Conclusion

In this research, we have examined the innovative topic of big data, which has recently gained lots of interest due to its perceived unprecedented opportunities and benefits. In the information era we are currently living in, voluminous varieties of high velocity data are being produced daily, and within them lay intrinsic details and patterns of hidden knowledge which should be extracted and utilized. Hence, big data analytics can be applied to leverage business change and enhance decision making, by applying advanced analytic techniques on big data, and revealing hidden insights and valuable knowledge.

Accordingly, the literature was reviewed in order to provide an analysis of the big data analytics concepts which are being researched, as well as their importance to decision making. Consequently, big data was discussed, as well as its characteristics and importance. Moreover, some of the big data analytics tools and methods in particular were examined. Thus, big data storage and management, as well as big data analytics processing were detailed. In addition, some of the different advanced data analytics techniques were further discussed.

By applying such analytics to big data, valuable information can be extracted and exploited to enhance decision making and support informed decisions. Consequently, some of the different areas where big data analytics can support and aid in decision making were examined. It was found that big data analytics can provide vast horizons of opportunities in various applications and areas, such as customer intelligence, fraud detection, and supply chain management. Additionally, its benefits can serve different sectors and industries, such as healthcare, retail, telecom, manufacturing, etc.

Accordingly, this research has provided the people and the organizations with examples of the various big data tools, methods, and technologies which can be applied. This gives users an idea of the necessary technologies required, as well as developers an idea of what they can do to provide more enhanced solutions for big data analytics in support of decision making. Thus, the support of big data analytics to decision making was depicted.

Finally, any new technology, if applied correctly can bring with it several potential benefits and innovations, let alone big data, which is a remarkable field with a bright future, if approached correctly. However, big data is very difficult to deal with. It requires proper storage, management, integration, federation, cleansing, processing, analyzing, etc. With all the problems faced with traditional data management, big data exponentially increases these difficulties due to additional volumes, velocities, and varieties of data and sources which have to be dealt with. Therefore, future research can focus on providing a roadmap or framework for big data management which can encompass the previously stated difficulties.

We believe that big data analytics is of great significance in this era of data overflow, and can provide unforeseen insights and benefits to decision makers in various areas. If properly exploited and applied, big data analytics has the potential to provide a basis for advancements, on the scientific, technological, and humanitarian levels.

References

1. Adams, M.N.: Perspectives on Data Mining. International Journal of Market Research 52(1), 11–19 (2010)
2. Asur, S., Huberman, B.A.: Predicting the Future with Social Media. In: ACM International Conference on Web Intelligence and Intelligent Agent Technology, vol. 1, pp. 492–499 (2010)
3. Bakshi, K.: Considerations for Big Data: Architecture and Approaches. In: Proceedings of the IEEE Aerospace Conference, pp. 1–7 (2012)
4. Cebr: Data equity, Unlocking the value of big data. in: SAS Reports, pp. 1–44 (2012)
5. Cohen, J., Dolan, B., Dunlap, M., Hellerstein, J.M., Welton, C.: MAD Skills: New Analysis Practices for Big Data. Proceedings of the ACM VLDB Endowment 2(2), 1481–1492 (2009)
6. Cuzzocrea, A., Song, I., Davis, K.C.: Analytics over Large-Scale Multidimensional Data: The Big Data Revolution! In: Proceedings of the ACM International Workshop on Data Warehousing and OLAP, pp. 101–104 (2011)
7. Economist Intelligence Unit: The Deciding Factor: Big Data & Decision Making. In: Capgemini Reports, pp. 1–24 (2012)

8. Elgendy, N.: Big Data Analytics in Support of the Decision Making Process. MSc Thesis, German University in Cairo, p. 164 (2013)

9. EMC: Data Science and Big Data Analytics. In: EMC Education Services, pp. 1–508 (2012)

10. He, Y., Lee, R., Huai, Y., Shao, Z., Jain, N., Zhang, X., Xu, Z.: RCFile: A Fast and Space-efficient Data Placement Structure in MapReduce-based Warehouse Systems. In: IEEE International Conference on Data Engineering (ICDE), pp. 1199–1208 (2011)

11. Herodotou, H., Lim, H., Luo, G., Borisov, N., Dong, L., Cetin, F.B., Babu, S.: Starfish: A Self-tuning System for Big Data Analytics. In: Proceedings of the Conference on Innovative Data Systems Research, pp. 261–272 (2011)

12. Kubick, W.R.: Big Data, Information and Meaning. In: Clinical Trial Insights, pp. 26–28 (2012)

13. Lee, R., Luo, T., Huai, Y., Wang, F., He, Y., Zhang, X.: Ysmart: Yet Another SQL-to-MapReduce Translator. In: IEEE International Conference on Distributed Computing Systems (ICDCS), pp. 25–36 (2011)

14. Manyika, J., Chui, M., Brown, B., Bughin, J., Dobbs, R., Roxburgh, C., Byers, A.H.: Big Data: The Next Frontier for Innovation, Competition, and Productivity. In: McKinsey Global Institute Reports, pp. 1–156 (2011)

15. Mouthami, K., Devi, K.N., Bhaskaran, V.M.: Sentiment Analysis and Classification Based on Textual Reviews. In: International Conference on Information Communication and Embedded Systems (ICICES), pp. 271–276 (2013)

16. Plattner, H., Zeier, A.: In-Memory Data Management: An Inflection Point for Enterprise Applications. Springer, Heidelberg (2011)

17. Russom, P.: Big Data Analytics. In: TDWI Best Practices Report, pp. 1–40 (2011)

18. Sanchez, D., Martin-Bautista, M.J., Blanco, I., Torre, C.: Text Knowledge Mining: An Alternative to Text Data Mining. In: IEEE International Conference on Data Mining Workshops, pp. 664–672 (2008)

19. Serrat, O.: Social Network Analysis. Knowledge Network Solutions 28, 1–4 (2009)

20. Shen, Z., Wei, J., Sundaresan, N., Ma, K.L.: Visual Analysis of Massive Web Session Data. In: Large Data Analysis and Visualization (LDAV), pp. 65–72 (2012)

21. Song, Z., Kusiak, A.: Optimizing Product Configurations with a Data Mining Approach. International Journal of Production Research 47(7), 1733–1751 (2009)

22. TechAmerica: Demystifying Big Data: A Practical Guide to Transforming the Business of Government. In: TechAmerica Reports, pp. 1–40 (2012)

23. Van der Valk, T., Gijsbers, G.: The Use of Social Network Analysis in Innovation Studies: Mapping Actors and Technologies. Innovation: Management, Policy & Practice 12(1), 5–17 (2010)

24. Zeng, D., Hsinchun, C., Lusch, R., Li, S.H.: Social Media Analytics and Intelligence. IEEE Intelligent Systems 25(6), 13–16 (2010)

25. Zhang, L., Stoffel, A., Behrisch, M., Mittelstadt, S., Schreck, T., Pompl, R., Weber, S., Last, H., Keim, D.: Visual Analytics for the Big Data Era—A Comparative Review of State-of-the-Art Commercial Systems. In: IEEE Conference on Visual Analytics Science and Technology (VAST), pp. 173–182 (2012)

Author Index